Lecture Notes on

CHERN-SIMONS-WITTEN THEORY

Sen Hu
Princeton University

Published by

World Scientific Publishing Co. Pte. Ltd.

P O Box 128, Farrer Road, Singapore 912805

USA office: Suite 1B, 1060 Main Street, River Edge, NJ 07661

UK office: 57 Shelton Street, Covent Garden, London WC2H 9HE

British Library Cataloguing-in-Publication Data
A catalogue record for this book is available from the British Library.

LECTURE NOTES ON CHERN-SIMONS-WITTEN THEORY

ISBN 981-02-3908-4
ISBN 981-02-3909-2 (pbk)

This book is printed on acid-free paper.

Printed in Singapore by Uto-Print

Lecture Notes on

CHERN-SIMONS-WITTEN THEORY

To my parents
To Bo, Maomao and Amy

Preface

More than 15 years ago, Vaughn Jones discovered an elegant construction of a new polynomial invariant for knots in three-dimensional space. Jones's discovery has been generalized in many directions and a variety of new definitions have been found.

Mathematically rigorous approaches to the Jones polynomial and its generalizations tend to be combinatorial in nature. These definitions often lack manifest three-dimensional symmetry, but this symmetry can sometimes be proved by showing invariance under an appropriate set of "moves".

There is also an alternative approach to this subject, discovered by "physical" methods a few years after Vaughn Jones's initial work. In this approach, the basic object of study is a three-dimensional quantum gauge theory in which the action is the Chern-Simons invariant of a connection. Knot invariants and three-manifold invariants can be defined in this theory, at a physical level of rigor. Using the Feynman path integral, the various invariants can be described in a way that manifest the full three-dimensional symmetry. (Indeed, Feynman originally introduced the path integral for a very similar reason – to make manifest the full relativistic symmetry of Quantum Electrodynamics.) On the other hand, by using the relation of the path integral to the Hamiltonian formalism, the combinatorial recipes that are more familiar mathematically can be deduced from this more invariant starting point.

But this is a difficult road for mathematicians to pursue. Success depends on putting on a rigorous basis the requisite quantum field theory techniques. This is not likely to be easy.

Quantum field theory, in which the quantum concepts are applied to fields and not just to particles, was in many ways the greatest and most difficult achievement in twentieth century physics. It is the basis for most of our present-day understanding of nature. But it is a hard subject that has developed in fits and starts and that despite valiant attempts is still largely out of reach mathematically. The gauge theory approach to the Jones polynomial and its generalizations places them in this central, yet as of now, mathematically inaccessible part of the physical and mathematical world.

What can mathematicians gain by trying anyway, or at least by learning something of what physicists have to say, even if it cannot yet be fully justified mathematically? The combinatorial definitions of the knot and three-manifold invariants are beautiful, but they are only one side of the story. Quantum field theory exposes a relation of these same invariants to gauge theory on the one hand, to conformal field theory and stable bundles on Riemann surfaces on the other hand, as well as to other mathematical theories like Donaldson theory and the theory of affine Lie algebras that also have a natural quantum field theory setting. Knowing all sides of the story, or as much as we can learn, is worthy in itself and may well be necessary for understanding applications of the knot and three-manifold invariants.

I hope that the present volume by Sen Hu (which is based in part on lectures I gave at Princeton University in the spring of 1989) will help make this subject more accessible to curious mathematicians. Hopefully, the explanations given here will help mathematical readers (who may also want to consult Michael Atiyah's book *The Physics and Geometry of Knots*) develop a wider understanding of this subject and its relations to physics. And perhaps it will impel some to help develop a more complete mathematical exposition of this subject than is now possible.

By E. Witten
July 22, 2000

Contents

Chapter 1

Examples of Quantizations

1.1 Quantization of $\mathbf{R}^2$

1.1.1 *Classical mechanics*

There are several equivalent formulations of classical mechanics, namely Newtonian, Lagrangian and Hamiltonian formalisms.

In Newtonian formalism N particles $q_1, ..., q_N$ with masses $m_1, ..., m_N$ attracting each other are governed by

$$m_i q_i^{''} = -\sum_{j \neq i} \frac{\partial}{\partial q_i} \left(-\frac{m_i m_j}{|q_i - q_j|} \right), i = 1, ..., N. \tag{1.1}$$

For $N = 2$ this is exactly soluble and it was used to explain several laws of Kepler. Lagrange reformulated the above equations into a variational problem:

$$\delta \int_a^b L dt = 0,$$

where L is a function of $q_1, ..., q_N, q_1^{'}, ..., q_N^{'}$,

$$L = \sum_i \frac{1}{2} m_i q_i^{'2} - \sum_{i \neq j} \frac{m_i m_j}{|q_i - q_j|}.$$

The Euler-Lagrange equation

$$\frac{d}{dt}\left(\frac{\partial L}{\partial q_i'}\right) = \frac{\partial L}{\partial q_i}, i = 1, ..., N \tag{1.2}$$

gives the Newton equations (1.1).

If we introduce new variables $p_i = \frac{\partial L}{\partial q_i'}$ and let $H = \Sigma_i p_i q_i - L$ be a function of $q_1, ..., q_N, p_1, ..., p_N$, then Euler-Lagrange equations can be re-written as

$$\frac{dp_i}{dt} = \frac{\partial L}{\partial q_i} = -\frac{\partial H}{\partial q_i},$$

$$\frac{dq_i}{dt} = \frac{p_i}{m_i} = \frac{\partial H}{\partial p_i}.$$

A remarkable property is that if

$$\Phi : (q_1, ..., q_N, p_1, ..., p_N) \rightarrow (Q_1, ..., Q_N, P_1, ..., P_N)$$

is a transformation induced by

$$\Sigma_i P_i dQ_i - \Sigma_i p_i dq_i = dS,$$

or Φ preserves the two form $\omega = \Sigma_i dp_i \wedge dq_i$, then for the new coordinates $Q_1, ..., Q_N, P_1, ..., P_N$, the equations have the same form.

In its abstract form classical mechanics can be formulated as follows:

Canonical formalism : Let (M^{2n}, ω) be a symplectic manifold. For each $H \in C^\infty(M^{2n}, \mathbf{R})$ one can associate a Hamiltonian vector field X_H on M^{2n}, such that

$$dH(.) = \omega(X_H, .).$$

The flows generated by X_H are called Hamiltonian flows.

In the beginning of the 20th century scientists tried to use a model of the solar system to describe atomic structures with some successes. However experiments forced them to abandon the classical picture and adopted the quantum picture.

Classically we use space-time $(q_1, q_2, q_3, q_4) \in \mathbf{R}^4$ as the basic variables and each point corresponds to an event. In quantum mechanics one uses the space of probabilities on space time as the basic variables and the observables are Hermitian operators on the space of probabilities. This is the most striking idea introduced by Heisenberg. There are several ways of quantization. The Feynman path integral corresponds to the Lagrange formalism. Corresponding to Hamiltonian formalism we have geometric quantization. In the following we will describe geometric quantization of $(\mathbf{R}^{2n}, \omega)$.

1.1.2 *Symplectic method*

Let $\mathbf{R}^2 = T^*\mathbf{R}^1$ be a symplectic manifold with the symplectic form $\omega = dx \wedge dp$ with coordinates (x, p).

Here is their quantization: take the Hilbert space $\mathcal{H} = L^2(\mathbf{R})$ with the inner product $(\phi, \psi) = \int \phi(x)\psi(x^{'})dxdx^{'}$. We want to construct a map

$$C^\infty(\mathbf{R}^2, \mathbf{C}) \to Her(L^2(\mathbf{R}), L^2(\mathbf{R}))$$

Here $Her(L^2(\mathbf{R}), L^2(\mathbf{R}))$ denote the space of Hermitian operators. There are two ways of realizations.

(A): $\mathcal{H} = L^2(\mathbf{R})$, consisting of functions of $x, \psi(x)$.

$$x : \psi \to x\psi, \psi \in L^2(\mathbf{R}),$$

$$p : \psi \to -i\frac{d\psi}{dx}.$$

One can verify that $[p, x] = -i$, here $[,]$ is the Poisson bracket induced from the symplectic form ω.

(B): $\mathcal{H}^{'} = L^2(\mathbf{R}^{'})$, consisting of functions of $p, \phi(p)$.

$$p : \phi \to p\phi,$$

$$x : \phi \to i\frac{d\phi}{dp}.$$

(A) and (B) are equivalent via Fourier transformation

$$T : \mathcal{H} \to \mathcal{H}',$$

$$\psi(x) \to (T\psi)(p) = \int \frac{1}{\sqrt{2\pi}} e^{ipx} \psi(x) dx.$$

Let us consider quantization of harmonic oscillators. In classical mechanics, harmonic oscillators is generated by the Hamiltonian

$$H(x,p) = \frac{x^2 + p^2}{2}$$

with respect to the 0 symplectic form $\omega = dx \wedge dp$.

In quantum mechanics we have the same Hamiltonian. Let

$$a = \frac{p - ix}{\sqrt{2}}, a^* = \frac{p + ix}{\sqrt{2}},$$

a^* is a adjoint of a. Then we have

$$H(x,p) = \frac{x^2 + p^2}{2} = a^* a + \frac{1}{2},$$

So $H \geq \frac{1}{2}$. It is easy to verify that

$$[a, a^*] = aa^* - a^* a = 1,$$

$$[H, a] = Ha - aH = -a,$$

$$[H, a^*] = Ha^* - a^* H = a^*.$$

So if ψ is an eigen-function of H, then $a\psi, a^*\psi$ are also eigen-functions of H, and

$$H(a\psi) = (\lambda - 1)a\psi,$$

$$H(a^*\psi) = (\lambda + 1)a^*\psi.$$

We call a the annilation operator and a^* the creation operator.

If χ is a ground state , i.e. $H\chi = \frac{1}{2}\chi$, then

$$\chi, a^*\chi, (a^*)^2\chi, \ldots$$

gives a basis of $\mathcal{H}$ with eigenvalues $\frac{1}{2}, \frac{3}{2}, \frac{5}{2}, \ldots$.

For example, $\chi = e^{-\frac{1}{2}x^2}$ is a ground state, then $a^{*n}\chi$ will be

$$\chi_n(x) = \text{const.}(-i\frac{d}{dx} + ix)^n e^{-\frac{1}{2}x^2} = H_n(x)e^{-\frac{1}{2}x^2},$$

where $H_n(x)$ are Hermite polynomials.

Classically under Lie bracket $\{,\}$, symplectic Lie algebra is represented by quadratic forms x^2, p^2, and $\frac{1}{2}(xp+px)$. This gives representation of the Lie algebra $sl(2, \mathbf{R})$ by

$$\frac{-ix^2}{2} \to \begin{pmatrix} 0 & 0 \\ 1 & 0 \end{pmatrix},$$

$$\frac{-ip^2}{2} \to \begin{pmatrix} 0 & -1 \\ 0 & 0 \end{pmatrix},$$

$$-\frac{i}{2}(xp + px) \to \begin{pmatrix} 1 & 0 \\ 0 & -1 \end{pmatrix}.$$

We can decompose the Hilbert space $\mathcal{H}$ into

$$\mathcal{H} = \mathcal{H}_1 \oplus \mathcal{H}_2,$$

$$\mathcal{H}_1 = \{(a^{*l}\chi, l \text{ even}\},$$

$$\mathcal{H}_2 = \{(a^{*l}\chi, l \text{ odd}\}.$$

1.1.3 *Holomorphic method*

There is a geometric quantization of $(\mathbf{R}^2, \omega)$ by a holomorphic method which will be more useful to us. We consider $\mathbf{R}^2 = \mathbf{C}$. Let

$$z = \frac{1}{\sqrt{2}}(p + ix), \bar{z} = \frac{1}{\sqrt{2}}(p - ix),$$

$$\omega = dx \wedge dp = -id\bar{z} \wedge dz.$$

So ω is a form of type $(1, 1)$. Now consider the space

$$\mathcal{H} = \{\text{holomorphic functions on } \mathbf{C}\}.$$

Let z act on $\mathcal{H}$ by multiplication and $\bar{z}$ act by $\frac{\partial}{\partial z}$. We define the inner product on $\mathcal{H}$ to be

$$(f, g) = \int_{\mathbf{C}} dz d\bar{z} e^{-\bar{z}z} \overline{f(z)} g(z),$$

then $\frac{\partial}{\partial z}$ is the conjugate of z.

The space $\mathcal{H}$ can be described in another way, i.e.

$$\mathcal{H} = \Gamma(\mathbf{C}, \mathcal{L}),$$

where $\mathcal{L} \to \mathbf{C}$ is the trivial complex line bundle and a section $s(z)$ of $\mathcal{L}$ is the same as a holomorphic function on $\mathcal{C}$.

We define a metric on $\Gamma(\mathbf{C}, \mathcal{L})$ by

$$||s||^2_{\mathcal{L}} = e^{-\bar{z}z}|s(z)|^2.$$

Then $(\mathcal{L}, ||.||)$ is a Hermitian line bundle. The metric we defined can be characterized by the property that it is compatible with a unitary connection whose curvature is the $(1, 1)$ form ω.

Here compatibility means

$$d(||s||^2_{\mathcal{L}}) = (s, Ds) + (Ds, s),$$

where D is the covariant derivative. To compute the curvature we pick a holomorphic section s, then

$$F = i\bar{\partial}\partial \log ||s||^2_{\mathcal{L}}.$$

Let us take $s = 1$, then $||s||^2_{\mathcal{L}} = e^{-\bar{z}z}$, so

$$F = i\bar{\partial}\partial \log ||s||^2_{\mathcal{L}} = i\bar{\partial}\partial(-\bar{z}z) = -id\bar{z}dz = \omega.$$

Let $\mathcal{H} = \Gamma_{L^2}(\mathbf{C}, \mathcal{L})$. We now construct another quantization as follows: z acts on $\mathcal{H}$ by multiplication, $\bar{z}$ acts on $\mathcal{H}$ as $\frac{\partial}{\partial z}$. 1 is annihilated by $\frac{\partial}{\partial z}$ and $\{z^n.1\}_{n=0,1,2,\ldots}$ forms a basis of $\mathcal{H}$.

Remark: This quantization depends on the choice of a complex structure. It is subtle to see how it varies as the complex structure varies.

1.2 Holomorphic representation of symplectic quotients and its quantization

In this chapter we will describe holomorphic representation of symplectic quotients and its quantizations. The novelty is that some symplectic quotients admit holomorphic description via geometric invariant theory and it is very convenient to quantize it in the holomorphic setting.

1.2.1 *An example of circle action*

Consider $\mathbf{R}^{2n} = \mathbf{C}^n$ with the standard symplectic structure $\omega = \sum_{i=1}^n dx_i \wedge dy_i = i\sum_{i=1}^n dz_i \wedge d\bar{z}_i$, or $\omega = -id\alpha, \alpha = \sum_{i=1}^n \bar{z}_i dz_i$.

Let us consider a circle action of $U(1)$ on $\mathbf{C}^n$ by

$$z_i \to e^{i\theta} z_i, i = 1, 2, ..., n, e^{i\theta} \in U(1).$$

This defines a homomorphism:

$$\rho : U(1) \to \mathrm{Diff}^{\omega}(\mathbf{R}^{2n}),$$

where $\mathrm{Diff}^{\omega}(\mathbf{R}^{2n})$ stands for diffeomorphisms of $\mathbf{R}^{2n}$ preserving the symplectic form ω.

If we take a derivative of the above map we would have the moment map:

$$d\rho : \mathrm{g} \to \mathrm{Symp}^{\omega}(\mathbf{R}^{2n}),$$

where $\mathbf{g}$ is the complex Lie algebra $\mathbf{C}$ of $U(1)$. $\mathrm{Symp}^{\omega}(\mathbf{R}^{2n})$ is the space of symplectic vector fields, i.e. vector fields of the form $(-H_y, H_x)$ and $H : \mathbf{R}^{2n} \to \mathbf{R}$ is a Hamiltonian function. In this case the symplectic vector field is:

$$V = \sum_i (z_i \frac{\partial}{\partial z_i} - \bar{z}_i \frac{\partial}{\partial \bar{z}_i}).$$

It generates a flow

$$\frac{dz_i}{dt} = z_i, \frac{d\bar{z}_i}{dt} = -\bar{z}_i,$$

and $z_i \bar{z}_i$ is a first integral.

The moment map is:

$$\mu(z) = \sum_i (z_i \bar{z}_i - 1)$$

via $d\mu = i_V \omega$. The inverse image of μ at a regular value gives a symplectic quotient of the $U(1)$ action.

$$\mathbf{C}^n // S^1 = \mu^{-1}(0)/S^1 = \{(z_1, ..., z_n) | \sum_i z_i \bar{z}_i = 1\}/S^1 = \mathbf{CP}^{n-1}.$$

On the other hand, we may consider

$$S^1 = U(1) \subset GL(1) = \mathbf{C}^*.$$

And we consider the complexified action:

$$z_i \to \lambda z_i, \lambda \in \mathbf{C}^*.$$

We see that

$$\mathbf{C}^n//S^1 = (\mathbf{C}^n)^{s.s}/\mathbf{C}^*,$$

where $(\mathbf{C}^n)^{s.s}$ are the set of semi-stable points of the $\mathbf{C}^*$ action on $\mathbf{C}^n$ consisting of non-zero vectors of $\mathbf{C}^n.z$ is a semi-stable point if and only if $\sum \bar{z}_i z_i$ is bounded away from zero along the $\mathbf{C}^*$ orbit $\mathbf{C}^*.z$.

1.2.2 *Moment map of symplectic actions*

The construction above applies for much general cases. To proceed further let us consider symplectic action on a complex manifold.

Let (M^{2n}, ω) be a symplectic manifold and let a Lie group G acts on M^{2n} preserving the symplectic form ω. So we have a homomorphism:

$$\Phi : G \to \mathrm{Diff}_\omega(M^{2n}),$$

where $\mathrm{Diff}_\omega(M^{2n})$ is the space of diffeomorphisms of M^{2n} preserving ω, i.e.

$$\Phi(g)^*(\omega) = \omega, g \in G.$$

By differentiating Φ we have

$$d\Phi : \mathbf{g} \to \mathrm{Symp}^\omega(M^{2n}).$$

Again $\mathrm{Symp}^\omega(M^{2n})$ is the space of symplectic vector fields which can be identified with the space of Hamiltonian vector fields, i.e. $i_\omega H$, with a smooth function $H : M^{2n} \to \mathbf{R}$. Here $i_\omega H$ is obtained by

$$\omega(., i_\omega H) = dH(.).$$

From this we define a one form:

$$\bar{J} : TM^{2n} \to \mathbf{g}^*,$$

$$< \bar{J}.v_x, \xi >= \omega_x(d\Phi(\xi), v_x),$$

for every $\xi \in \mathrm{g}$.

In the case that $H^2(\mathrm{g}) = 0$, or $H^1(M^{2n}) = 0$, we can always integrate the above one form and get a function:

$$J : M^{2n} \to \mathrm{g}^*$$

such that $dJ = \bar{J}.J$ is usually called the moment map of the action.

Here are more examples:

1) Rotational action

We have the symplectic manifold $M = T^*V, V = \mathbf{R}^3$, with the symplectic form $\omega = d\alpha, \alpha = pdq$. $SO(3)$ acts on V by rotation.

One can calculate the moment functions by

$$H_\alpha(x) = \alpha(\frac{d}{dt}|_{t=0} g^t x),$$

and they are actually the moment functions $M_1 = p_2 q_3 - p_3 q_2, M_2 = p_3 q_1 - p_1 q_3, M_3 = p_1 q_2 - p_2 q_1$.

2) Natural action of $U(n+1)$ on $\mathbf{CP}^n$

Here the action is given by

$$\rho : U(n+1) \times \mathbf{CP}^n \to \mathbf{CP}^n,$$

$$(u, x) \to u.x.$$

One can verify that the following function defines the moment map:

$$\mu : \mathbf{CP}^n \to \mathcal{U}(n+1)^*$$

$$\mu(x).a = (2\pi i||x^*||^2)^{-1} \bar{x}^{*t} a x^*,$$

$$a \in \mathcal{U}(n+1), x^* \in \mathbf{C}^{n+1} - \{0\}.$$

3) Lie group K acts on a complex algebraic manifold X

Let a Lie group K act on a complex manifold X which is embedded in $\mathbf{CP}^n$. The action is the induced action of $U(n+1)$ on $\mathbf{CP}^n$. Then the moment map is:

$$\mu : X \to \mathbf{k}^*,$$

$$\mu(x).a = (2\pi i||x^*||^2)^{-1}\bar{x}^{*t}\phi_x(a)x^*.$$

1.2.3 *Some geometric invariant theory*

The constructions above apply for much general cases via the Geometric Invariant Theory developed by Hilbert, Grothendick and Mumford.

Let X be a complex manifold which acts on by a compact group G. Let $\mathcal{L} \to X$ be an ample holomorphic line bundle, i.e. there exists an integer N such that X is embedded in $\mathbf{CP}^N$ by sections in $\Gamma(X, \mathcal{L}^N)$, via

$$x \to (s_1(x), ..., s_{N+1}(x)).$$

G action lifts to an action on an ample holomorphic line bundle. When G is compact then $G \subset U(N+1)$ for some N. Then we have $G_{\mathbf{C}}$ which acts on $\mathbf{CP}^N$ as well as on X. From the Geometric Invariant Theory we have

$$X//G = X^{s.s.}/G_{\mathbf{C}}. \tag{1.3}$$

Here $X^{s.s.}$ are the set of semi-stable points in X, i.e. those point z in X such that invariant polynomials are bounded away from zero on the orbit $G_{\mathbf{C}}z$.

Let us illustrate it by one more example. Consider the circle action on $(\mathbf{R}^{2n}, \omega)$ whose complexification is the action $\mathbf{C}^*$ which acts on $(\mathbf{C}^n, \omega)$ as

$$(z_1, ..., z_n) \to (\lambda z_1, ..., \lambda z_s, \lambda^{-1}z_{s+1}, ..., \lambda^{-1}z_n).$$

The invariant polynomials are $z_i z_j, i \le s, j > s$. $z \in \mathbf{C}^n$ is semi-stable if and only if those invariant polynomials are bounded away from zero on the orbit of $\mathbf{C}^*.z$. Or equivalently $\mathbf{C}^*.z$ is bounded away from zero, or $\sum_i \bar{z}_i z_i$

is bounded away from zero on the orbit of $\mathbf{C}^*.z$. So unstable points z are such that all $z_i = 0$, for $i \leq s$, or all $z_j = 0, j > s$.

The moment map of the above action is:

$$\mu : X \to \mathbf{g}^*, \mu(x) = \frac{x.\bar{x}}{||x||^2}, ||x||^2 = \sum \bar{z}_i z_i.$$

1.2.4 *Grassmanians*

Given a vector space V over $\mathbf{C}$ of dimension N, the Grassmanian $G(k, N; \mathbf{C})$ is defined as the space of k-dimensional subspaces in V. It can be repre-sentated by k linear independent vectors $e_1, ..., e_k \subset V$. Let $B \subset \mathbf{C}^{kN} = V \times V \times ... \times V$ be the space of k linear independent vectors. B is an open dense subset in $\mathbf{C}^{kN}.GL(k, \mathbf{C})$ acts on B.

$$G(k, N; \mathbf{C}) = B/GL(k, \mathbf{C}).$$

Let g be a Hermitian metric on V. On $V \times ... \times V = \mathbf{C}^{kN}$, we choose a basis

$$\phi^{is}, i = 1, ..., k, s = 1, ..., N.$$

Let $\omega = i \sum_{i,s} d\phi^{is} \wedge d\bar{\phi}_{is}$. Here $\phi_{is} = g_{ij} g_{sk} \phi^{jk}$. ω is invariant under the group action $U(k) \subset GL(k, \mathbf{C})$. In other words, $U(k)$ acts on $\mathbf{C}^{kN}$ symplectically. The moment map in this case is

$$\mu : \mathbf{C}^{kN} \to u(k)^*, (e_1, ..., e_k) \to \{(e_i, e_j) - \delta_{ij}\}.$$

So we have

$$\mu = 0 \Longleftrightarrow e_1, ..., e_k \text{ orthonormal}.$$

Here we arrive at another natural representation of Grassmanians

$$G(k, N; \mathbf{C}) = \mu^{-1}(0)/U(k).$$

1.2.5 *Calabi-Yau/Ginzburg-Landau correspondence*

The following example came from the paper by E. Witten, "Phases of $N = 2$ theories in two dimensions", Nuclear Physics B 403 (1993) 159-222. In this paper the $N = 2$ supersymmetric nonlinear σ-model in a special Calabi-Yau target space is reduced to the following problem in algebraic geometry.

Let us consider a hypersurface

$$V = \{(a_1, a_2, b_1, b_2) \in \mathbf{C}^4 | a_1 b_2 - a_2 b_2 = 0\}.$$

There is a natural $U(1)$ action on $\mathbf{C}^4$ which leaves V invariant,

$$a_i \to \lambda a_i, b_j \to \lambda^{-1} b_j, \lambda \in U(1).$$

$V//U(1)$ is a Calabi-Yau threefold because there exists a global holomorphic 3-form on it, i.e.

$$\Theta = da_1 \wedge da_2 \wedge db_1 \wedge db_2.$$

The moment map for the $U(1)$ action is

$$U(a_i, b_j) = (|a_1|^2 + |a_2|^2 - |b_1|^2 - |b_2|^2 - r)^2.$$

Here r reflects the possibility of adding a constant to the Hamiltonian. For $r > 0$, the set of semi-stable points is

$$Z_+ = (V' \cup V_1)/\mathbf{C}^*,$$

$$V' = \{(a_1, a_2, b_1, b_2) \in \mathbf{C}^4 | (|a_1|^2 + |a_2|^2)(|b_1|^2 + |b_2|^2) \neq 0\},$$

$$V_1 = \{(a_1, a_2, 0, 0) \in \mathbf{C}^4\}.$$

Z_+ can be considered as a fiber bundle over $\{(a_1, a_2, 0, 0) \in \mathbf{C}^4\}$, i.e. we have

$$\mathbf{C}^2 \to Z_+ \to \mathbf{CP}^1_a.$$

Zero sections are genus zero holomorphic curves.

For $r < 0$, the set of semi-stable points is

$$Z_- = (V' \cup V_2)/\mathbf{C}^*,$$

$$V_2 = \{(0, 0, b_1, b_2) \in \mathbf{C}^4\}.$$

It is also a fiber bundle

$$\mathbf{C}^2 \to Z_- \to \mathbf{CP}^1_b.$$

We see that there is a transition. The singular locus is switched from a-space to b-space. Such a transition is called a flop in algebraic geometry. Our point here is that Z_+ and Z_- came from the same space symplectically. So we expect the quantum theory the same for the two cases. This is the rationale of persistence of quantum theory of space time with respect to changing of topology in σ-models/ string theory.

1.2.6 *Quantization of symplectic quotients*

We have illustrated examples of quantization in Chapter 1. The constructions there applies for much general cases.

1) Quantization of the cotangent bundle

Let N be a manifold and T^*N be its cotangent bundle. On T^*N there is a natural symplectic structure, i.e. $\omega = d\alpha, \alpha = \sum_i p_i dq_i$, where $(q_1, ..., q_n)$ are coordinates for N and $(p_1, ..., p_n)$ are coordinates for fibers.

The Hilbert space for quantizing T^*N is

$$\mathcal{H} = L^2(N).$$

We map coordinates functions to operators on H as follows:

$$q_i : \psi \to q_i\psi,$$

$$p_i : \psi \to -i\frac{\partial\psi}{\partial q_i}, \psi \in \mathcal{H}.$$

2) Quantization of a Kahler manifold

Let (M^{2n}, ω) be a Kahler manifold with ω the Kahler form so it is naturally a symplectic manifold. We can construct as a holomorphic line bundle $\mathcal{L}$ over M^{2n} and a Hermitian metric $||.||$ on $\mathcal{L}$, such that the curvature of this metric is

$$F = -i\partial\bar{\partial}\log||s||^2 = -\omega.$$

Example: On $\mathbf{CP}^n$ we have the tautologically line bundle $\mathcal{L} \to \mathbf{CP}^n$. Let us consider the metric

$$||s||^2_{\mathcal{L}} = e^{-\bar{z}z}|s|^2,$$

$|s|^2$ is the Euclidean metric on the fiber. Then the curvature of this metric is:

$$F = -i\partial\bar{\partial}\log||s||^2 = i\partial\bar{\partial}(-\bar{z}z) = -id\bar{z}dz = \omega.$$

In general for a Kahler manifold we can choose

$$\mathcal{L} = \det H^*(E) = (\Lambda^{Top}H^0(M^{2n}, E) \times \Lambda^{Top}(H^1(M^{2n}, E))^{-1}.$$

To quantize the Kahler manifold we take the Hilbert space

$$\mathcal{H} = \Gamma(M^{2n}, \mathcal{L}),$$

that is the space of holomorphic sections of the chosen line bundle. There are natural operators such as multiplication by a section or differentiation by a covariant derivative.

3) Quantization of a symplectic quotient

Let X be a complex manifold which acts on by a compact group G holomorphically. From the Geometric Invariant Theory we have

$$X//G = X^{s.s.}/G_{\mathbf{C}}.$$

This suggests we quantize the quotient space by quantizing X and then passing through its quotients. Namely we choose a Hilbert space as:

$$\mathcal{H} = \Gamma(X/G, \mathcal{L}) = \Gamma(X^{s.s.}/G_{\mathbf{C}}, \mathcal{L}) = \Gamma(X^{s.s.}, \mathcal{L})^{G_{\mathbf{C}}}.$$

The Hilbert space consists of $G_{\mathbf{C}}$ part of the space of holomorphic sections of the bundle $\mathcal{L} \to X^{s.s}$. The corresponding operators to coordinate functions are the same as before.

Remark: For applications we take $X = \mathbf{R}^{2n}$ and G compact Lie groups. One may see [Jeffrey-Kirwan], [Meinrenken], [Tian-Zhang], for different approaches and recent developments.

Chapter 2

Classical Solutions of Gauge Field Theory

2.1 Moduli space of classical solutions of Chern-Simons action

Our main interest is in gauge field theory. The discovery of a gauge invariant Lagrangian by Yang and Mills is a breakthrough in gauge field theory. For three-manifold, Chern-Simons Lagrangian is also invariant under gauge transformation up to an integer. It is important to emphasize that the basic variables are gauge equivalent fields. In this section we will describe the moduli space of classical solutions for Chern-Simons action. It is the space of flat connections with some useful additional structures.

2.1.1 *Symplectic reduction of gauge fields over a Riemann surface*

Readers may also refer to Atiyah and Bott's paper "Philo. Trans. of Royal Soc. London", A308 (1982) 523, for more details.

Let Σ be a Riemann surface, i.e. a topological surface with a choice of complex structure $J : T\Sigma \to T\Sigma, J^2 = -Id$. Let $E \to \Sigma$ be a bundle with structure group G. Let $\mathcal{A}$ be the space of connections A on the G bundle E. A is a g-valued one form, i.e. $A \in \Gamma(\Sigma, E \otimes \mathrm{g})$, g is the Lie algebra of G. The gauge group $\mathcal{G} = \{$ maps $\Sigma \to G\}$ acts on $\mathcal{A}$, the space of connections, via

$$(g, A) \to dgg^{-1} + gAg^{-1}.$$

Infinitesimally we have u, a g-valued 0-form on Σ, and

$$Du := du + [A, u], A(\epsilon) = A - \epsilon Du.$$

There is a natural symplectic structure on $\mathcal{A}$ given by

$$\omega = \frac{k}{2\pi} \int_{\Sigma} \mathrm{Tr}(\delta A \wedge \delta A).$$

Here δ is the exterior derivative on $\mathcal{A}$. Du is a symplectic vector field.
Remark: $\mathcal{G}$ acts symplectically on $\mathcal{A}$.
Problem: Find the moment map of the $\mathcal{G}$ action on $\mathcal{A}$.
It amounts to solve the equation

$$\delta\mu(u) = i_{Du}\omega.$$

$$i_{Du}\omega = \frac{k}{4\pi} \int_{\Sigma} \mathrm{Tr}((-Du) \wedge \delta A)$$

$$= \frac{k}{4\pi} \int_{\Sigma} \mathrm{Tr}(u \wedge D(\delta A)).$$

For a one-parameter family of connections $A_\epsilon = A + \epsilon B$, we have the curvature

$$F_\epsilon = dA_\epsilon + A_\epsilon \wedge A_\epsilon$$

$$= F + \epsilon DB + O(\epsilon^2),$$

So $\delta F = D\delta A$.
Hence,

$$i_{Du}\omega = \frac{k}{4\pi} \int_{\Sigma} \mathrm{Tr}(u\Delta F)$$

$$= \frac{k}{4\pi}\delta \int_\Sigma \mathrm{Tr}(uF).$$

This means that the curvature $F = dA + A \wedge A$ is the moment map.

By symplectic reduction we get the symplectic quotient, which sits in the space of gauge equivalent field , $\mu^{-1}(0)/G$, i.e. the space of flat connections. For each of such a connection A and a loop C in Σ, we have holonomy $\int_C \mathrm{Tr}A$. This determines a representation $\rho : \pi_1(\Sigma) \to G$. It is not difficult to see that the representation also determines A up to gauge transformation. Hence we have

$$\mu^{-1}(0)/G = \text{ Hom } (\pi_1(\Sigma), G)/G,$$

where G acts by conjugation.

The space of gauge equivalent flat connections inherits the natural symplectic structure from $\mathcal{A}$, i.e. $\omega = \int_\Sigma \mathrm{Tr}(\delta A \wedge \delta A)$. It can be described in terms of cohomology as follows.

The tangent space of $\mu^{-1}(0)$ is $\mathrm{Ker}D$. Here $Du = du + [A, u]$ is the covariant derivative. We have then complexes

$$\Omega^0(\Sigma; E \otimes \mathbf{g}) \xrightarrow{D} \Omega^1(\Sigma; E \otimes \mathbf{g}) \xrightarrow{D} \Omega^2(\Sigma; E \otimes \mathbf{g})$$

The tangent space of $\mathcal{M} = \mu^{-1}(0)/G$ is

$$T\mathcal{M} = \text{ Ker}D/ \text{ Image}D = H^1(\Sigma; E \otimes \mathbf{g}).$$

And one can verify easily that

$$\omega(\alpha_1, \alpha_2) = \int_\Sigma \mathrm{Tr}(\alpha_1 \wedge \alpha_2)$$

is a symplectic form on $\mathcal{M}$, i.e. it only depends on the homology classes of α_1 and α_2.

2.1.2 *Chern-Simons action on a three manifold*

Let M^3 be a three manifold, G a Lie group, $E \to M$ a principal G-bundle. Let A be a connection on E, i.e. $A \in \Omega^1(M^3, E \otimes \mathbf{g})$. We consider the

Chern-Simons action

$$L = \frac{k}{4\pi}\int_M \mathrm{Tr}(A\wedge dA + \frac{2}{3}A\wedge A\wedge A), k\in \mathbf{Z}^+,$$

taking values in $\mathbf{R}/2\pi\mathbf{Z}$.

Here G is a Lie group with Lie algebra $\mathbf{g}$ on which we endow with a non-degenerate quadratic form,

$$(a,b) = \mathrm{Tr}ab.$$

Again the gauge group is

$$\mathcal{G} = \{\text{ maps } M\to G\}$$

each of them corresponds to a change of trivialization of the bundle $E\to M$.

The gauge group $\mathcal{G}$ acts on the space of connections $\mathcal{A}$ via

$$(g,A)\to gAg^{-1} + dgg^{-1}.$$

One can verify that Chern-Simons functional is invariant under the action of gauge transformations up to an integer which corresponds to the homotopy class of the gauge transformation. So it is a functional over the space of gauge equivalent connections valued at $\mathbf{R}/2\pi\mathbf{Z}$.

Remark: Pick a four manifold B with $\partial B = M$, extend E, A over B, then we have

$$\int_{\partial B} \text{Chern-Simons} = \frac{k}{4\pi}\int_B \mathrm{Tr}(F\wedge F).$$

If we choose a different B' with $\partial B' = M$ and consider $X = B\cup(-B')$, here $-B'$ is a manifold with reversed orientation. Then we have

$$\frac{1}{4\pi}\int_X \mathrm{Tr}(F\wedge F) = \frac{1}{4\pi}(\int_B \mathrm{Tr}(F\wedge F) - \int_{B'} \mathrm{Tr}(F\wedge F))$$

is an integer representing the first Pontryagin class of X.

The classical solution of Chern-Simons functional are flat connections on M^3. It can be seen as follows.

Take a one-parameter family of connections $A_\epsilon = A + \epsilon B$, where B is also $\mathbf{g}-$ valued 1-form.

$$L_{C.S.}(A_\epsilon) = \frac{k}{4\pi}\int \mathrm{Tr}((A+\epsilon B)(dA+\epsilon dB)+\frac{2}{3}(A+\epsilon B)\wedge(A+\epsilon B)\wedge(A+\epsilon B))$$

$$= L_{C.S}(A) + \epsilon\frac{k}{4\pi}\int \mathrm{Tr}(B\wedge F) + O(\epsilon^2).$$

Since B are arbitrary we have $F = 0$.

One knew that flat connections are determined by their holonomy, i.e. by representations $\pi_1(M) \to G$. So we have

critical points/gauge transformations $<==>$ $\mathrm{Hom}(\pi_1(M), G)/G$.

Now consider a special 3 manifold $M^3 = \Sigma \times \mathbf{R}$, where Σ is a surface. We write connection A on $E \to M$ as $A = a + A_0 dt, A_0$ is a 0-form.

Then we have:

1) $\frac{\delta L}{\delta A_0} = 0 <==> da + a\wedge a = 0$, or $a(t)$ is a flat connection on $\Sigma\times\{t\}$.

2) $\frac{\delta L}{\delta a} = 0, ==> \frac{da}{dt} + DA_0 = 0$, this means that up to gauge transformation, what happens at $t \neq 0$ is determined by the connection at $t = 0$.

Note that the infinitesimal gauge transformations are:

$$A_0 \to A_0 + \epsilon\frac{Du}{dt}, \frac{Du}{Dt} = \frac{du}{dt} + [A_0, u],$$

$$a \to a + \epsilon Du.$$

So after getting rid of A_0 and t we have that the gauge group becomes a gauge group on the surface and the moduli space becomes

$$\mathcal{M} = \mu^{-1}(0)/G\ ,$$

where $\mu : \mathcal{A} \to \mathbf{g}^*, a \to F = da + a\wedge a$ is the moment map. On $\mathcal{A}$, there is a natural symplectic structure above.

2.2 Maxwell equations and Yang-Mills equations

2.2.1 *Maxwell equations*

The following is the well known Maxwell equations:

$$\nabla B = 0, \nabla E = 0,$$

$$\frac{\partial B}{\partial t} + \nabla \times E = 0, \frac{\partial E}{\partial t} - \nabla \times B = 0.$$

There is a duality to this equation, i.e. it does not change with respect to the transformation $(E, B) \to (-B, E)$. This duality leads to an important recent development of Seiberg-Witten theory.

There are also continuous symmetries. It is H. Weyl who made this apparent in terms of gauge symmetry. It is actually a $U(1)$ gauge theory. To achieve this, we write E and B as components of a matrix:

$$F = (F^{\mu\nu}) = \begin{pmatrix} 0 & -E^1 & -E^2 & -E^3 \\ E^1 & 0 & -B^3 & B^2 \\ E^2 & B^3 & 0 & -B^1 \\ E^3 & -B^2 & B^1 & 0 \end{pmatrix}$$

$F = F^{\mu\nu} dx_\mu \wedge dx_\nu$ is a 2-form. If we write $E = (E^1 dx_1 + E^2 dx_2 + E^3 dx_3) \wedge dt, B = (B^1 dx_1 + B^2 dx_2 + B^3 dx_3) \wedge dt$, then $F = E + *B$. Here, $*$ is the Hodge operator. $*(dt \wedge dx_1) = dx_2 \wedge dx_3, *(dx_1 \wedge dx_2) = dt \wedge dx_3, \ldots$ It is easy to see that $*^2 = Id$. So we have $*F = *E + B$. Then it is easy to verify that the Maxwell equations can be expressed as

$$dF = 0, d(*F) = 0.$$

Since $dF = 0$ we have $F = dA$ if the domain is simply connected. A is usually called a vector potential or a gauge. Apparently if we change A to $A + d\Lambda, F$ does not change. We see that the Maxwell equations can be expressed as a $U(1)$ gauge theory. We consider L a $U(1)$ principle bundle over a domain in $\mathbf{R}^4$. Gauge field is $U(1)$ Lie algebra, i.e. $\mathbf{R}$, valued one-form. Gauge transformations are $g = e^{i\Lambda}, \Lambda \in \mathbf{R}$ when the bundle is trivial, in general they are sections of the bundle. g acts on A by $A^{'} = gAg^{-1} - ig^{-1}dg$.

Interesting questions arise if the bundle L is non-trivial. This motiviates P. Dirac to search for monopoles. Monopoles cannot exist if the bundle is trivial. If the bundle is non-trivial we would have a non-trivial invariant

$$c_1(L) = \int_{\Sigma} F$$

for any surface $\Sigma \subset \mathbf{R}^4$. This is the first Chern class which can be intepreted as flux in physics.

2.2.2 *Yang-Mills equations*

Very interesting things happen when we change the above gauge group $U(1)$ to a non-Abelian group, say $SU(2)$. Yang and Mills made the change in the mid-fifties to explain iso-spin in physics. It turns out that the setting, which express symmetry very well, is most important.

Let $E \to M$ be a principle bundle with a Lie group G as its fiber. Let A be a connection, i.e. $A = \sum_a \omega_a T^a$ be a g valued one form where T^a are generators of the Lie algebra g and ω_a are one forms. A gauge transformation g is a G valued zero form, i.e. a section of the bundle. g acts on the space of connections by $A^{'} = gAg^{-1} + gdg^{-1}$.

Given a connection A, we consider the covariant derivative $D = d + A$. We have $F = DA = dA + A \wedge A$ is a two-form, it is the curvature of A. A gauge transformation g transforms A into $A^{'}$ whose curvature is $F^{'} = gFg^{-1}$. We say that F transforms covariantly. Another remarkable property of F is the Bianchi identity $DF = 0$.

Principle bundle arises naturally in geometry and in physics. In both cases one needs to choose local coordinates. Different coordinates are related by gauge transformations, and geometric or physical quantities should be gauge invariant or covariant. The following functional on the space of connections, Yang-Mills functional, is gauge invariant,

$$L_{YM} = \frac{1}{2\pi} \int_M \mathrm{Tr}(F \wedge *F).$$

Here $*$ is the usually Hodge operator induced by a metric on M. Since F transforms covariantly, L_{YM} is invariant under a gauge transformation.

The Euler-Langrange equations of Yang-Mills are:

$$DF = 0, D(*F) = 0.$$

Since $DF = 0$ is an identity, the non-trivial equations are $D(*F) = 0$.

When the dimension of M is four, if a connection A satisfies $*F = \pm F$, then it automatically satisfies the Yang-Mills equations. Such solutions are called self-dual and anti-self-dual instantons. It plays a key role in Donaldson's theory of differentiable structures of four manifold.

Another remarkable property is that F can be used to describe topological properties although it is only a local data. For example, the first Chern class $c_1(E) = \frac{1}{2\pi i}\mathrm{Tr}F \in H^2(M, \mathbf{Z})$ is independent of the representation of connections. This can be seen as follows.

Given a connection A, consider its variations $A^{'} = A + \delta A$ whose curvature is given by $F^{'} = F + D\delta A$. Notice that $\mathrm{Tr}D\delta A = d\mathrm{Tr}\delta A$. For any two cycle Σ, we have

$$\int_{\Sigma} \frac{1}{2\pi i}\mathrm{Tr}F^{'} = \int_{\Sigma} \frac{1}{2\pi i}\mathrm{Tr}(F + D\delta A) = \int_{\Sigma} \frac{1}{2\pi i}\mathrm{Tr}F.$$

By the same reasoning plus Bianchi identity, the second Chern class $c_2(E) = \frac{1}{2\pi i}\mathrm{Tr}(F \wedge F)$ is also a topological invariant. It is also the Pontryagin class of E. The second Chern class $c_2(E)$ for an instanton is called the instanton number which is a topological invariant.

The Yang-Mills functional can be viewed as a functional over the space of gauge equivalent field. Bott showed that the Yang-Mills functional is a perfect Morse function over the space of gauge equivalent field, and from this he determined cohomologies of the classifying space of the space of gauge transformations [Atiyah-Bott(1982)].

It is interesting to note that the Chern-Simons action $C.S.$ of ∂M and the second Chern class or Pontryagin class $c_2(E)$ of M are related by

$$dC.S. = c_2(E).$$

2.3 Vector bundle, Chern classes and Chern-Weil theory

2.3.1 *Vector bundle and connection*

Definition: (Vector bundle) Let V be a vector space, M a manifold. We say $E \to M$ is a vector bundle, if $E = \cup_\alpha U_\alpha \times V$, where $M = \cup_\alpha U_\alpha$, and if $U_\alpha \cap U_\beta \neq \phi$, we identify $U_\alpha \times V$ with $U_\beta \times V$ by a transition function

$$g_{\alpha\beta} : U_\alpha \cap U_\beta \to \mathrm{Hom}(V, V),$$

with $(x_\alpha, v_\alpha) = (y_\beta, g_{\alpha\beta} v_\beta), x_\alpha = y_\beta$.

Here $g_{\alpha\beta}$ is a cocycle. It satisfies:

1)$g_{\alpha\alpha} = 1$,

2)$g_{\alpha\beta} = g_{\alpha\gamma} g_{\gamma\beta}$, for any point in $U_\alpha \cap U_\beta \cap U_\gamma$.

We have a canonical projection $\pi : E \to M, (x, v) \to x$. There are many natural examples of vector bundles such as tagent bundle, cotangent bundle of a manifold and their tensor products.

Definition: (Connection) The notion of connection generalizes the concept of directional derivative. We knew that the derivative $\frac{d}{dx}$ acts on the space of functions. For a vector bundle the space of functions is generalized to the space of sections

$$\Gamma(E) = \{s : M \to E | \pi s = Id\}.$$

Then a connection can be defined as a linear operator:

$$D : \Gamma(E) \to \Gamma(E \otimes T^*(M)).$$

It satisfies the Lebnitz rule:

$$D(fs) = df \otimes s + fDs.$$

In local coordinates, let $\{e_i\}$ be a basis of sections so that every section s can be representated by $s = \Sigma_i s_i e_i$. Let

$$De_i = \Sigma_j \theta_{ij} e_j,$$

where the connection matrix $A = (\theta_{ij})$ is represented by a matrix of one forms.

If we choose a different set of local coordinates, they are related by a gauge transformation $g : M \to G \subset \mathrm{Hom}(V, V)$. The connection A is then represented by $A^{'} = gAg^{-1} + dgg^{-1}$, where dgg^{-1} is the Maurer-Cartan form of G. For each connection A, we define covariant derivative $D_A = d{+}A$. It is easy to see that covariant derivative transforms covariantly under a gauge transformation: $D_{A'} = gD_Ag^{-1}$.

2.3.2 *Curvature, Chern classes and Chern-Weil theory*

Definition: (Curvature) Curvature of a connection A is defined as $\Omega = D_A^2 = dA + A \wedge A$. It enjoys two important properties:

1) Bianchi identity: $D\Omega = 0$.

2) For a different set of local coordinates differed by a gauge transformation g, we have $\Omega^{'} = g\Omega g^{-1}$.

Definition: (Chern classes and Chern forms) From the Bianchi identity, we can verify that $c_i(E) = \frac{1}{2\pi i}\mathrm{Tr}\Omega^i$ is a closed form in $H^{2i}(M, \mathbf{R})$. It is called the $i - th$ Chern class of E. Chern classes can also be defined by: $\det(I + \frac{1}{2\pi i}\lambda\Omega) = \Sigma_i \lambda^i c_i(E)$.

If we modify A to $A + \delta A$, the curvature is changed to $\Omega + D_A\delta A$. By this and the Bianchi identity it is easy to verify that the Chern classes $c_i(E) = \frac{1}{2\pi i}\mathrm{Tr}\Omega^i$ as cohomology classes are independent of choices of A. This is the beautiful **Chern-Weil theory**.

Chapter 3

Quantization of Chern-Simons Action

3.1 Introduction

From the last chapter we see that the classical solutions of Yang-Mills over a Riemann surface or Chern-Simons over $\Sigma \times \mathbf{R}$ is the moduli space

$$\mathcal{M}_\Sigma^G = \mathrm{Hom}(\pi_1(\Sigma), G)/G.$$

Here G is any compact semi-simple Lie group with an invariant quadratic form $(,) = \frac{k}{4\pi}\mathrm{Tr}, k \in \mathbf{Z}^+$ on its Lie algebra $\mathbf{g}$. There is a natural symplectic form

$$\omega = \int_\Sigma (\delta A, \delta A)$$

over the space of connections. So $\mathcal{M}_\Sigma^G$ inherits this symplectic structure by passing to homology classes of connections. So it is together with ω forms a symplectic space. We want to quantize such a space, i.e. to construct a Hilbert space associated to $\mathcal{M}_\Sigma^G$ or Σ,

$$\Sigma \to \mathcal{H}_\Sigma(G, (,)).$$

In geometric quantization we first construct a line bundle $\mathcal{L} \to \mathcal{M}$. The Hilbert space will be the space of holomorphic sections, i.e. $\Gamma(\mathcal{M}, \mathcal{L}^k), k \in \mathbf{Z}^+$. The construction which we will give at first depends on picking up a complex structure J on Σ. However one can show that such a construction is independent of complex structures by constructing a projective flat connection on the bundle $\mathcal{H}_\Sigma(G,(,)) \to$ Teich, the space of complex structures over the surface. We can then identify different Hilbert spaces by using this flat connection. We may take advantage of independent of complex structures. We can decompose Σ into pants by choosing a set of maximally disjoint simple closed curves C on Σ. For each pant decomposition we will construct a Hilbert space $\mathcal{H}$ by assigning each loop around puncture an irreducible representation of G. Then the Hilbert space $\mathcal{H}$ for Σ can be constructed from (by taking the tensor product) Hilbert spaces for the collection of pants (conformal blocks).

In the following we will realize such a construction for a list of Lie groups. It turns out that it often gives many interesting applications out of those constructions.

1) $\mathbf{R}$

2) $S^1 = \mathbf{R}/2\pi\mathbf{Z}$

This construction leads to classical theta functions on a Jacobi variety.

3) $T^*G = \mathbf{g}^* \times G, \mathbf{g}$ is the Lie algebra of G regarded as an Abelian group acted on by G. The Lie algebra of T^*G is

$$L = \text{Lie}(T^*G) = \mathbf{g}_{ab} + \mathbf{g}_G$$

$\{a,b\} \in L, a, b, \in \mathbf{g}, (\{a,b\},\{a',b'\}) = b'(a) - b(a')$. This construction leads to Reidermeister torsion or η invariants. If we replace T^*G by Super T^*G, it gives Casson invariant [Witten-Casson].

4) Compact semi-simple Lie group, e.g. $G = SU(2)$.

This construction leads to a series of knot invariants including Jones polynomials.

5) Non-compact semi-simple Lie group, e.g. $SL(2,\mathbf{R}), SL(2,\mathbf{C})$ with an invariant quadratic form which are integral and non-degenerate. This construction seems to have interesting connections to Thurston's geometrization program.

3.2 Some formal discussions on quantization

Let us first consider quantization of the Chern-Simons action informally. It turns out the following informal consideration is very illuminating and it can be made rigorous mathematically from the work of [Axelrod-DellaPietra-

Witten].

We consider quantization for a special three manifold $M^3 = \Sigma \times [0,1]$. The phase space is now the space of connections $\mathcal{A}$ on Σ. $\mathcal{A}$ is an affine symplectic space with symplectic form $\omega(\alpha,\beta) = \int_\Sigma \mathrm{Tr}(\alpha \wedge \beta)$. Gauge group $\mathcal{G}$ acts on $\mathcal{A}$ symplectically.

Remark: We also learned in the last chapter that the space of classical solutions $\mathcal{M}$ is the space of flat connections. For each flat connection A we have a covariant derivative $d_A = d + A$. The tangent space of $\mathcal{M}$ is the space of first cohomology $H^1_{d_A}(\Sigma, E \otimes \mathbf{g})$.$\mathcal{M}$ is also a symplectic variety with respect to the symplectic form ω above and one can easily check that ω only depend on d_A-cohomology classes. This way we push the symplectic form down to a symplectic form on the symplectic quotient.

As we have seen in Chapter One it is pretty easy to quantize an affine symplectic space. We use holomorphic quantization here. To do this we need to have a complex structure on $\mathcal{A}$. There are natural complex structures coming from a choice of complex structure $J : T\Sigma \to T\Sigma, J^2 = -Id$, of the underlying surface Σ. The induced complex structure on $\mathcal{A}$ is nothing but to claim the $(1,0)$ part of the connection $A = A_z dz + A_{\bar{z}} d\bar{z}$ to be holomorphic. With this complex structure the quantization is quite simple. The Hilbert space is then the space of holomorphic sections of the trivial line bundle $\mathcal{L} = \mathcal{A} \times \mathbf{C}$.

We wish to say a few more words on the bundle $\mathcal{L}$. We define functional derivative or a connection on $\mathcal{A}$ to be

$$\frac{D}{DA_z} = \frac{\delta}{\delta A_z} - \frac{k}{4\pi} A_{\bar{z}},$$

$$\frac{D}{DA_{\bar{z}}} = \frac{\delta}{\delta A_{\bar{z}}} + \frac{k}{4\pi} A_z.$$

One can check that

$$[\frac{D}{DA_{\bar{z}}}(z), \frac{D}{DA_w}(w)] = \frac{k}{2\pi}\delta(z,w).$$

So the curvature for this connection is $-i\omega$. And formally the first Chern class of the line bundle $\mathcal{L}$ is $i\omega$. We post the first condition on the big Hilbert space $\Psi \in \Gamma(\mathcal{A}, \mathcal{L}^{\otimes k})$ as,

$$\frac{D}{DA_{\bar{z}}}\Psi = 0.$$

Recall that what we want to quantize is the space of gauge equivalent connections $\mathcal{A}/\mathcal{G}$. The idea is to quantize affine space of the space of connections and then select the gauge invariant part, i.e.

$$\Gamma_{hol}(\mathcal{A}/\mathcal{G}, \mathcal{L}^{\otimes k}) = \Gamma_{hol}(\mathcal{A}, \mathcal{L}^{\otimes k})^{\mathcal{G}}.$$

We also have $\mathcal{G}_{\mathbf{C}}$, the group of complexified gauge group, acts on $\mathcal{A}$. A holomorphic section which is invariant under $\mathcal{G}$ is also invariant under $\mathcal{G}_{\mathbf{C}}$. So formally we have

$$\Gamma_{hol}(\mathcal{A}, \mathcal{L}^{\otimes k})^{\mathcal{G}}$$

$$= \Gamma_{hol}(\mathcal{A}, \mathcal{L}^{\otimes k})^{\mathcal{G}_{\mathbf{C}}}$$

$$= \Gamma_{hol}(\mathcal{A}/\mathcal{G}_{\mathbf{C}}, \mathcal{L}^{\otimes k}).$$

The space $\mathcal{A}/\mathcal{G}_{\mathbf{C}}$ is nothing but the space of holomorphic bundles over the surface Σ. According to a theorem of Narasimhan-Seshadri, the space of stable holomorphic $G_{\mathbf{C}}$ bundles is the same as the space of flat G connections on Σ. This explains why the Hilbert space is $\Gamma_{hol}(\mathcal{M}, \mathcal{L}^{\otimes k})$. This is a well defined Hilbert space even though other Hilbert spaces above are ill defined.

To get $\Gamma_{hol}(\mathcal{A}, \mathcal{L}^{\otimes k})^{\mathcal{G}}$ we need to get the condition to select gauge invariant sections. It is given by

$$(D_{\bar{z}}\frac{\delta}{\delta A^a_z} - \frac{k}{4\pi}F_{z\bar{z}a})(\Psi) = 0.$$

This follows from the fact that F is the moment map for the symplectic action of the gauge group on the space of connections. The above gives a rough idea on how to construct the Hilbert space. We will give rigorous construction of the Hilbert space in the following section. There is an importan issue of how the Hilbert space varies as complex structures varies. We shall address this problem in section 3.5.

3.3 Pre-quantization

3.3.1 $\mathcal{M}$ *as a complex variety*

The moduli space is

$$\mathcal{M} = \text{Hom}(\pi_1(\Sigma), G)/G.$$

For each $A \in \mathcal{M}$, define its covariant derivative $d_A = d + A$. Since $d_A^2 = 0$ for a flat connection, it induces a cohomogy on de Rham complexes. Gauge fields are then better expressed in terms of cohomology classes. The tangent space of $\mathcal{M}$ is $T\mathcal{M} = H^1_{d_A}(\Sigma, E \otimes \mathrm{g})$. It's dimension is

$$\dim \mathcal{M} = \dim H^1_A(\Sigma; E \otimes \mathrm{g}) = (2g-2)\dim G.$$

Pick a complex structure J on Σ, we make $\mathcal{M}$ a complex variety. If we express connection A on Σ as $A = A_z dz + A_{\bar{z}} d\bar{z}, \nabla = \nabla^{(1,0)} + \nabla^{(0,1)}, \nabla^{(1,0)} f = \Sigma_i \frac{\partial f}{\partial z_i} dz_i, \nabla^{(0,1)} f = \Sigma_i \frac{\partial f}{\partial \bar{z}_i} d\bar{z}_i$, then we see that the complex structure on $\mathcal{M}$ is nothing but claiming dz_i the holomorphic part and $d\bar{z}_i$ the anti-holomorphic part. J induces a $*$ operator on $\Omega^1(\Sigma, E\otimes \mathrm{g}), *^2 = Id$. Define $I : T\mathcal{M} \hookleftarrow, I\alpha = - * \alpha. I$ is an integrable complex structure.

The symplectic structure ω is compatible with the complex structure $J, \omega(\alpha, I\alpha) = -\int_\Sigma \text{Tr}(\alpha \wedge *\alpha) \geq 0$ and $g(\alpha, \beta) = \omega(\alpha, I\beta)$ is a Kahler metric $(\omega(\alpha, I\beta) = \omega(I\alpha, \beta))$ with the Kahler form $\omega \in \Omega^{1,1}(\mathcal{M})$.

There is also another explicit description of $\mathcal{M}$ as a complex manifold by a theorem of Narasimhan and Seshadri [Narasimhan-Seshadri]. Their theorem states that a holomorphic vector bundle E on a compact Riemann surface Σ is stable if and only if it arises from a unitary flat connection. Here stability means that for each holomorphic sub-bundle $U \subset E$,

$$\frac{\deg U}{\text{rank } U} < \frac{\deg E}{\text{rank } E}.$$

Hence the moduli space is the space of holomorphic bundles over Σ with respect to a choosen complex structure. This space has a natural complex structure. The above theorem of Narasimhan and Seshadri has been generalized to Kahler manifolds in high dimensions by Donaldson-Uhlenbeck-Yau [Uhlenbeck-Yau].

The above results can be viewed as an infinite dimensional version of geometric invariant theory. It is very interesting that even the existence of Calabi-Yau metrics also admits an interpretation of geometric invariant theory, see [Donaldson].

3.3.2 *Quillen's determinant bundle on $\mathcal{M}$ and the Laplacian*

In geometric quantization, the next step is to construct a line bundle $\mathcal{L} \to \mathcal{M}$. This is given by Quillen's determinant bundle $\mathcal{L}_A = \det d_A$.

Let $\mathcal{M} = \mathrm{Hom}(\pi_1(\Sigma, G))/G, A \in \mathcal{M}$ be a flat connection on a bundle E over a Riemann surface $\Sigma, \mathrm{rank}(E) = r, \deg E = d$. The determinant bundle is defined as a bundle $\mathcal{L}$ over $\mathcal{M}$. At A, we define the fiber to be

$$\mathcal{L} = \lambda(\mathrm{ker} d_A)^* \otimes \lambda(\mathrm{coker} d_A),$$

where $\lambda(V)$ is the highest exterior power of V.

Let $d_A : \Omega^{0,0}(E) \to \Omega^{0,1}(E), \Omega^{0,0}(E) = \mathrm{ker} d_A \oplus V_0, \Omega^{0,1}(E) = \mathrm{coker} d_A \oplus V_1$. The Laplacian can be defined as $\Delta = d_A d_A^* + d_A^* d_A : V_0 \to V_0$. It is natural to define $\det \Delta = \prod \lambda_i = (\frac{\det V_0}{\det V_1})^2$.

Let us consider $\zeta(s) = \sum_n \frac{1}{\lambda_n^s}$, then $\exp(-\zeta'(0)) = \prod_n \lambda_n$.

Quillen also defines a metric ([Quillen])

$$||\sigma_{d_A}||^2 = \det{}_\zeta(d_A^* d_A) = \exp(-\zeta'(0)).$$

He shows that the curvature of the determinant bundle with respect to this metric is equal to the Kähler form on $\mathcal{M}$.

3.4 Some Lie groups

3.4.1 $G = \mathbf{R}$

The moduli space is

$$\mathcal{M}_\Sigma^{\mathbf{R}} = \mathrm{Hom}(\pi_1(\Sigma), \mathbf{R}) = H^1(\Sigma, \mathbf{R})$$

with the standard quadratic form ω. It can be quantized by the usual methods as described in chapter 1.

3.4.2 $G = S^1 = \mathbf{R}/2\pi\mathbf{Z}$

The moduli space is

$$\mathcal{M}_\Sigma^{S^1} = \mathrm{Hom}(\pi_1(\Sigma), S^1) = \mathbf{R}^{2g}/\mathbf{Z}^{2g} = \mathrm{Jac}(\Sigma).$$

It is the Jacobi variety. Given a complex structure J on Σ, it induces a complex structure on $\mathrm{Jac}(\Sigma)$. It is actually given by a basis of holomorphic differentials $\omega_1, \omega_2, ..., \omega_g$ and $\mathrm{Jac}(\Sigma) = \mathbf{C}^g/\mathbf{Z}^{2g}$ is represented by

$$\left(\int \omega_1, ..., \int \omega_g\right) \in \mathbf{C}^g$$

divided by its periods

$$\left(\int_{\alpha_1} \omega_1, ..., \int_{\alpha_g} \omega_g; \int_{\beta_1} \omega_1, ..., \int_{\beta_g} \omega_g\right) \in \mathbf{Z}^{2g},$$

where $\alpha_1, ..., \alpha_g; \beta_1, ..., \beta_g$ is a basis of $H_1(\Sigma, \mathbf{Z})$.

To quantize $\mathrm{Jac}(\Sigma)$, we first construct a line bundle $\mathcal{L} \to \mathrm{Jac}(\Sigma)$ with a natural connection A on $\mathcal{L}$ whose curvature is $-\omega$, *i.e.*, $dA = -\omega$.

Here ω is a 2-form on $H^1(\Sigma, \mathbf{R})$, the tangent space of the moduli space $\mathrm{Jac}(\Sigma)$,

$$\omega(\alpha, \beta) = \int_\Sigma \alpha \wedge \beta.$$

Note also that

$$\omega \in H^{1,1}(\mathrm{Jac}, \mathbf{Z}) \to H^2(\mathrm{Jac}, \mathbf{R})$$

is the first Chern class of $\mathcal{L}$.

To quantize $\mathrm{Jac}(\Sigma)$, we consider the Hilbert space

$$\mathcal{H}_\Sigma = \Gamma(\mathrm{Jac}(\Sigma), \mathcal{L}^r)$$

$$= \text{Space of level r } \Theta - \text{functions}.$$

In $H^1(\Sigma, \mathbf{R})$, pick a set of "a-cycles", $a_1, ..., a_g$ a set of non-intersecting simple closed curves representing $H^1(\Sigma, \mathbf{R})$ and its dual basis "b-cycles", $b_1, ..., b_g$. We can choose a basis of holomorphic differentials $\omega_1, ..., \omega_g$, such that

$$\int_{a_i} \omega_j = \delta_{i,j},$$

$$\int_{b_i} \omega_j = \Omega_{i,j}.$$

And Θ function has the following form:

$$\Theta_m(u, \Omega) = \sum_{l \in \Omega^g, l = m \bmod k} \exp\left(\frac{\pi i}{k} < l, Zl > + 2\pi i < l, u >\right).$$

3.4.3 T^*G

In this case the moduli space is

$$\mathcal{M}_\Sigma^{T^*G} = \mathrm{Hom}(\pi_1(\Sigma), T^*G)/G,$$

$$= T^*(\mathrm{Hom}(\pi_1(\Sigma), G)/G) = T^*(\mathcal{M}_\Sigma^G).$$

As we discussed earlier, the Hilbert space is

$$\mathcal{H}_\Sigma^{T^*G,(,)} = L^2_{\text{half density}}(\mathcal{M}_\Sigma^G).$$

The corresponding topological quantum field theory gives Reidemeister torsion and η-invariant. The Super T^*G theory gives Casson invariant [Witten-Casson].

3.5 Compact Lie groups, $G = SU(2)$

Now we construct geometric quantization of Chern-Simons theory for compact Lie groups, e.g. $SU(2)$. We will describe the Hilbert space for small genus and/or with very few punctures and calculate their dimension. This is very useful to explain some key facts for Jones polynomials.

3.5.1 *Genus one*

Consider a simple closed curve γ in $\mathbf{T}^2$ and after we collapse it we have a sphere with a double point. If we forget about the double point we have $\mathbf{CP}^1$.

Let $\epsilon \to \mathbf{CP}^1$ be a plane bundle, we have $c_1(\epsilon) = 0$. So

$$\epsilon = \mathcal{L}^k + \mathcal{L}^{-k}, \deg \mathcal{L} = 1.$$

Now let $\mathbf{CP}^1 = (\mathbf{CP}^1 - \{0\}) \cup (\mathbf{CP}^1 - \{\infty\}) = \mathcal{U}_1 \cup \mathcal{U}_2$. In $\mathcal{U}_1 \cap \mathcal{U}_2$ we take

$$f = \begin{pmatrix} z^k & 0 \\ 0 & z^{-k} \end{pmatrix}$$

as the transition map, this defines a bundle on $\mathbf{CP}^1$. We will construct a linear map $\phi : \epsilon_{P_1} \to \epsilon_{P_2}$ to identify them. Up to conjugacy we have

$$\phi = \begin{pmatrix} a & 0 \\ 0 & a^{-1} \end{pmatrix}.$$

So the moduli space is

$$\mathcal{M}_\Sigma^{s.s.} = T/W = \mathbf{C}^*/\mathbf{Z}_2,$$

where $T \subset G$ is a maximal torus and W is the Weyl group.

To quantize it, the Hilbert space is

$$\mathbf{C}^\infty(T/W).$$

It is generated by the following basis:

$$f_1 = 1,$$

$$f_2 = a + a^{-1},$$

$$f_3 = a^2 + 1 + a^{-2},$$

$$f_4 = a^3 + a^2 + a^{-2} + a^{-3},$$

......

where $f_n =$ {characters of the n dimensional representations of $SU(2)$ and the first n+1 functions have order of pole $\leq n$ at ∞}.

So we have

$$\mathcal{H}_\Sigma = \{\text{the first k+1 characters of } SU(2)\}.$$

In other words, quantization for a given k is equivalent to form a space consisting of meromorphic functions on T/W with poles of order $\leq k$ and it is equivalent to form the space of representations of $SU(2)$ of dimension $\leq k+1$.

We have constructed a map such that for a choice of a cycle a and a representation R of $SU(2)$ of dimension $\leq k+1$, we have $\psi_{(a,R)} \in \mathcal{H}_\Sigma$.

3.5.2 *Riemann sphere with punctures*

In the study of representation of compact Lie groups, there is an interesting geometric construction by the Borel-Weil-Bott theorem. It is in the same spirit of quantization. They found that the space G/T is naturally a symplectic space, where $T \subset G$ is the maximal torus, with a natural symplectic structure ω. They also found that G/T is isomorphic to $G_{\mathbf{C}}/B_{\mathbf{C}}$, the complexified group module a Borel subgroup, which is similar as we have seen in geometric invariant theory. They constructed a natural line bundle whose first Chern class is ω. Then one can form a Hilbert space, $V = H^0(G/T, L^k)$, the space of holomorphic sections of the line bundle. The group G acts as right translation and it also acts on the space of sections. This gives a linear representation of G. They showed that all irreducible representation arise from such a construction. This is a very useful fact for us in the quantization of Chern-Simons theory.

Now we consider the case of Riemann sphere with several punctures. At each puncture P_i, we associate a representation R_i. To each R_i, there is a definite conjugate class $u_i \in G$. The moduli space is:

$$\mathcal{M}_{(G,k;P_i,R_i)} = \mathrm{Hom}(\pi_1(\bar{\mathbf{C}} - \cup P_i), G)/G$$

with monodromy around P_i in conjugacy class u_i.

There is a similar result of Narasiham-Seshedri:

$\mathcal{M}_{G,k,\Sigma,P_i,R_i} = \{\epsilon|$ holomorphic $G_{\mathbf{C}}$ bundle on Σ, plus a reduction of structure group of ϵ to $B_{\mathbf{C}}$ at each $P_i\}$

Then the Hilbert space is:

$$\Gamma((\Pi_i(G_{\mathbf{C}}/B_{\mathbf{C}})_i/G_{\mathbf{C}}, \mathcal{L})$$

$$= \Gamma(\Pi_i(G_{\mathbf{C}}/B_{\mathbf{C}})_i, \mathcal{L})^{G_{\mathbf{C}}} = (R_1 \otimes \ldots \otimes R_s)^G.$$

The last equality is from the above Borel-Weil-Bott theorem.

Let us consider several examples. We want to determine the dimension of $\mathcal{H}_{(\Sigma;P_1,R_1;\ldots;P_s,R_s)}$ for small s.

a) $s = 0$

There is only one stable bundle $\epsilon \to S^2$. $\mathcal{M}(\Sigma)$ is a point consisting of the trivial representation. And $\dim \Gamma(pt, \mathcal{L}) = 1$. So $\dim \mathcal{H}_\Sigma = 1$.

b) $s = 1$

$$\mathcal{H}(\Sigma; P, Q) = R^G.$$

So $\dim \mathcal{H} = 1$, if R is trivial, and 0 otherwise.
c) $s = 2$

$$\dim \mathcal{H}(\Sigma; P_1, R_1, P_2, R_2) = \dim(R_1 \times R_2)^G = 1,$$

if $R_2 =$ dual of R_1 and 0 otherwise.

d) $s = 3$
It is the conformal block $N_{ij}^k = (R_i \otimes R_j \otimes R_k)^G$.
e) $s = 4$

$$\dim \mathcal{H}(\Sigma; P_1, R_1, P_2, R_2, P_3, R_3, P_4, R_4) = \dim(R \otimes R \otimes \bar{R} \otimes \bar{R})^G = 2.$$

Here R is the fundamental representation of $SU(N)$. It can be seen as follows. There is a canonical decomposition of $R \otimes R = E_1 \oplus E_2, E_1$ is the symmetric part, generated by $\frac{1}{2}(e_i \otimes e_j + e_j \otimes e_i)$, E_2 is the anti-symmetric part, generated by $\frac{1}{2}(e_i \otimes e_j - e_j \otimes e_i)$. So we have $R \otimes R \otimes \bar{R} \otimes \bar{R} = (E_1 \oplus E_2) \otimes (\bar{E_1} \oplus \bar{E_2})$. The $SU(2)$ invariant subspace is then generated by $E_1 \otimes \bar{E}_1$ and $E_1 \otimes \bar{E}_1$, hence it has dimension 2. This is the key to explain skein relations in knot polynomials.

3.5.3 *Higher genus Riemann surface*

Now we consider the case of higher genus Riemann surface. We want to quantize $\mathcal{M}_\Sigma^{flat}$ or what amounts to be the same, to quantize $\mathcal{M}_\Sigma^{\text{hol. stable}}$.

Let us first consider the case of genus two surface. We choose 3 simple closed curves on Σ. After we collapse them we get two copies of three punctured spheres. A bundle ϵ over Σ is then decomposed into two bundles $\mathcal{S}$ and $\mathcal{T}$. We think of gluing them together with maps $\phi_i : \mathcal{S}_{P_i} \to \mathcal{T}_{Q_i}$. For ϕ_i we also have equivalent relation ϕ_i with $u\phi_i v, u, v \in SL(2, \mathbf{C})$.

Roughly speaking, the Hilbert space is:

$$\mathcal{H}_\Sigma = \Gamma(G_L/G \times G \times G/G_R, \mathcal{L})$$

$$= \Gamma(G \times G \times G, \mathcal{L})^{G_L \times G_R}.$$

Holomorphic function of $G_{\mathbf{C}}$ as a representation of $G_L \times G_R$ is

$$\mathrm{Hol}(\mathrm{Fun}(G_{\mathbf{C}})) = \otimes_{R_i} R_i \otimes \bar{R}_i.$$

Here R_i runs over all equivalent classes of finite representations of G. And we have

$$\mathrm{Hol\ Fun}(G_1 \times G_2 \times G_3) = \otimes_{R_i,R_j,R_k}(R_i \otimes R_j \otimes R_k) \otimes (\bar{R}_i \otimes \bar{R}_j \otimes \bar{R}_k)$$

and we want it be $G_L \times G_R$ invariant.

The Hilbert space for the case of three punctured sphere Σ_0 is $\mathcal{H}_{\Sigma_0} = (R_i \otimes R_j \otimes R_k)^G = N_{ijk}$.

For genus two surface we have

$$\mathcal{H}_\Sigma = \oplus_{i,j,k}((R_i \otimes R_j \otimes R_k)^{G_L} \otimes (R_i \otimes R_j \otimes R_k)^{G_R}$$

$$= \oplus_{ijk} N_{ijk} \otimes N_{\bar{i}\bar{j}\bar{k}}.$$

The sum runs over all first $k+1$ representations of $SU(2)$.

For higher genus surface, we decompose the surface Σ into three punctured spheres by using a maximal set of simple closed curves. For each decomposition of the Riemann surface, there is a dual graph Γ. We assign to each edge c a weight $f(c) \in \mathcal{P}_k = \{0, 1/2, ..., k/2\}$. It obeys three conditions:

1)$|f(c_1) - f(c_2)| \leq f(c_3) \leq f(c_1) + f(c_2)$,

2)$f(c_1) + f(c_2) + f(c_3) \in \mathbf{Z}$,

3)$f(c_1) + f(c_2) + f(c_3) \leq k$.

Each admissible weight is an element of basis of the Hilbert space.

3.5.4 *Relation with WZW model and conformal field theory*

We will explain briefly that the quantization of Chern-Simons over a disk led to a natural connection with WZW model and conformal field theory. We will see that the moduli space consists of loop group and hence it is naturally connected with representations of affine Lie algebra.

Now the surface is a disk. The moduli space is the space of flat connections. The gauge group consists of gauge transformations which is an identity on the boundary. Since the disk is simply connected, we can find a gauge transformation U such that after applying the gauge transformation we have the connection is zero. So we have $A^{'} = UAU^{-1} + dUU^{-1} = 0$. And the connection can be represented as $A_i = -U^{-1}\partial_i U$, for some $U : D \to G$. The moduli space for the case of disk is then the loop group module the G action by conjugation, i.e. $\mathcal{M} = LG/G$.

There is a canonical symplectic structure coming from its representation as a flat connection, $\omega = \int \mathrm{Tr}(\alpha \wedge \beta)$. One can also construct a line bundle $\mathcal{L}$ over LG/G and a connection on this bundle whose curvature is ω. To quantize, we have the Hilbert space $\mathcal{H} = \Gamma_{hol}(LG/G, \mathcal{L}^k)$.$LG$ acts on LG/G by right translation. Hence this gives a projective representation of the loop group. One might wonder why a version of infinite dimensional Borel-Weil-Bott theorem gives many interesting representations of loop groups as well as affine Lie algebras.

The Chern-Simons action reduces to:

$$L_{WZW} = \int_{\mathbf{R}\times D} \mathrm{Tr}(U^{-1}dU \wedge U^{-1}dU \wedge U^{-1}dU) + \int_{\mathbf{R}\times \partial D} \mathrm{Tr}(U^{-1}\partial_\phi U U^{-1}\partial_t U).$$

It is the chiral Wess-Zumino-Witten (WZW) action. The WZW model is a physical model naturally connected with conformal field theory. We thus suggest a relation between Chern-Simons theory in the bulk with the conformal field theory (CFT) in the boundary.

In the case of gauge group $G = SL(2, \mathbf{R}), SL(2, \mathbf{C})$, it gives AdS/CFT duality, amid the recent proposal in gauge theory/string duality. One should note that the heart of the problem is to prove that the quantization is independent of complex structures. Witten established this for complex gauge groups[Witten-non-compact].

3.6 Independence of complex structures

Since we are aiming at a topological Chern-Simons theory, it is important to see that the quantization of Chern-Simons theory only depends on topological data. In geometric quantization we first choose a complex structure J on Σ. The space of connections becomes a complex space with $(1,0)$ part of a connection considered as holomorphic. In the above we construct the Hilbert space by using holomorphic quantization. Now we shall consider how the Hilbert space varies as complex structure varies. It is a vector bundle over the space of complex structures. We shall construct a projectively flat connection for this bundle.

The space of connections is an infinite dimensional affine space. It is better to consider quantization of finite dimensional space first. Let (A, ω) be a symplectic affine space with ω a symplectic form. As we have seen in Chapter One we may quantize it by using holomorphic method. To do this we choose a complex structure on A. For example in choosing standard complex structure with standard symplectic form we may write $\omega = \Sigma dp_i \wedge dq_i = \Sigma dz_i \wedge d\bar{z}_i$. We see this gives a different polarization of the symplectic form. There is an natural line bundle $L \to A$ whose first Chern class is ω. The Hilbert space is then the space of holomorphic sections, $\Gamma_{hol}(A, L^k)$. It is a vector bundle over the space of complex structures on A.

A complex structure is also an almost complex structure $J : TM \to TM, J^2 = -Id$. Let δJ be a small deformation of J, then we have $(J + \delta J)^2 = -Id$. This is $J\delta J + \delta J J = 0$. Let δJ be a deformation of complex structure J.

The following is a connection on the Hilbert space vector bundle.

$$\delta^{\mathcal{H}} = \delta + \mathcal{O}^{up}, \mathcal{O}^{up} = \frac{1}{8} M^{\bar{i}\bar{j}} \nabla_{\bar{i}} \nabla_{\bar{j}}, M^{\bar{i}\bar{j}} = (\delta J \omega^{-1})^{\bar{i}\bar{j}},$$

$$[\nabla_i, \nabla_{\bar{j}}] = k\omega_{i\bar{j}}, [\nabla_i, \nabla_j] = 0, [\nabla_{\bar{i}}, \nabla_{\bar{j}}] = 0.$$

It is important to note that:

1) $\delta^{\mathcal{H}}$ preserves holomorphicity. This can be verified by showing that it commute with complex derivatives $\Delta_{\bar{i}}$.

2) $(\delta^{\mathcal{H}})^2 =$ central. This means that it is a projectively flat connection.

Next we consider more interesting case of an affine symplectic space with a Lie group G action which preserves the symplectic form. We then have symplectic reduction and the symplectic quotient is $M = \mu^{-1}(0)/G$, where μ is the moment map. In the case of a finite dimensional affine space we can push down objects like symplectic form, complex structure, natural line bundle, connection etc. from A to M. In other words those objects can be constructed equivariantly. And in particular we have the projective flat connection for the Hilbert space bundle over complex structures.

For the case of infinite dimensional affine space of connections we may do similar things except that the big Hilbert spaces we met are often ill defined. We have a very large gauge group acting on the space of connections. And the symplectic quotient is finite dimensional. It would be nice if we could push objects interested to us down from the infinite dimensional affine space to objects on the symplectic quotiens. And this was done in [Axelrod-DellaPietra-Witten].

We first consider the case of $U(1)$ gauge group. In this case the space of sections are nothing but Θ functions. We write Θ functions in the form:

$$\Theta_m(u,Z) = \sum_{l \in \mathbf{Z}^g, l=m \bmod k} \exp(\frac{\pi i}{k} < l, Zl > +2\pi i < l, u >).$$

It satisfies:

$$\frac{1}{k}\frac{\partial^2 \Theta_m}{\partial u_i \partial u_j} = 2\pi i(1+\delta_{ij})\frac{\partial \Theta_m}{\partial Z_{ij}}.$$

In other words, Θ_m are covariant constant sections of a connection over the complex structures on $\mathbf{R}^{2n} = GL(2n;\mathbf{R})/GL(n;\mathbf{C})$. This implies projective flatness of the bundle of the Hilbert space over the space of complex structures.

There is another case where we can write down the projective flat connection explicitely. In the case of a punctured sphere, the projective flat connection is given by the Knizhnik-Zamolodchikov equations:

$$(\frac{\partial}{\partial z_i} - \frac{1}{k+h}\Sigma_{j\neq i}\frac{T_{(i)}T_{(j)}}{z_i - z_j})\Psi = 0;$$

$$\frac{\partial}{\partial \bar{z}_i}\Psi = 0,$$

where $\Psi \in R_1 \otimes ... \otimes R_n \subset \Gamma(\mathcal{M}, \mathcal{L}^k)$ is a wave form. All differential operators acting on Ψ are commutative. In Sec. 4.1.1 we will have more explanations.

Now we outline our treatment for the general case.

Recall in Sec. 3.3 we constructed a natural line bundle $\mathcal{L}$ over the symplectic quotient as Quillen's determinant bundle. Let ∇ be the connection coming from Quillen metric whose curvature is ω, i.e. $\nabla_{\bar{i}}\nabla_j s - \nabla_j\nabla_{\bar{i}} s = k\omega_{ij}s$, where s is a section.

Recall that the Hilbert space is

$$\Gamma(\mathcal{M}, \mathcal{L}^{\otimes k}) = \Gamma(\mathcal{A}/\mathcal{G}_{\mathbf{C}}, \mathcal{L}^{\otimes k}) = \Gamma(\mathcal{A}, \mathcal{L}^{\otimes k})^{\mathcal{G}_{\mathbf{C}}}.$$

Formally we have the projective flat connection as

$$\delta^* = \delta + \int_\Sigma d^2 z \delta\rho_{z\bar{z}} \frac{\delta}{\delta A_{\bar{z}}} \frac{\delta}{\delta A_{\bar{z}}}.$$

Since we have a large gauge group acting on the space of affine connections we need to push down the connection into a connection on a line bundle on $\mathcal{A}/\mathcal{G}_{\mathbf{C}}$. For a nice derivation of the above connection and a complete proof that the connection is projectively flat please see [Axelrod-DellaPietra-Witten]. We shall write down their formular presently. Before doing that we shall explain how to describe deformations of a complex structure in our case.

The tangent space of the moduli space is

$$T_*\mathcal{M} = H^1_A(\Sigma; E \otimes \mathbf{g}) = T^{1,0} \oplus T^{0,1}.$$

An infinitesimal deformation of the complex structure I is

$$\dot{I}: T^{0,1} \to T^{1,0}, \dot{I}\, I + I\, \dot{I} = 0.$$

The deformation is given by

$$G: T^{1,0} \xrightarrow{\omega} T^{0,1} \xrightarrow{\dot{I}} T^{1,0},$$

$G(\alpha,\alpha) = \int_\Sigma \mathrm{Tr}(\alpha \wedge * \alpha), \mathrm{Tr}\alpha^2 \in H^0(\Sigma; K^2)$ is a holomorphic quadratic form. In other words, $\delta J = \sum_{i,j,k} G^{ij}\omega_{j\bar{k}} \frac{\partial}{\partial z_i} \otimes d\bar{z}_k$.

Finally the quantum connection is:

$$\delta^* = \delta + \frac{1}{2t}(\nabla_i \delta J^{ij} \nabla_j + \delta J^{ij}(\nabla_i \log H)\nabla_j + \frac{1}{2}\frac{k^*}{k^* + h}\delta \log H),$$

where δ is the derivation of complex structures, $H = \log \Delta$ and Witten et al proved that δ^* is a projective flat connection. One may recognize that the first term in the bracket is similar to the one in the finite dimensional case. The rest of the terms in the bracket are due to anomalies.

Remark: Given the above projectively flat connection the Hilbert space can be constructed as the space of covariant constant sections of the Hilbert space bundle over the space of complex structures. And mapping class group acts on this Hilbert space and this gives a projective representations of the mapping class group. We shall describe this action in Chapter 4.

Remark: There is a nice relation between Chern-Simons-Witten theory and Wess-Zumino-Witten theory because that they share the same form of projective flat connections. In WZW theory the connection has a nice physical origin, that is, it comes from the energy-momemtum tensor. For a nice treatment of WZW model and its relation with Chern-Simons theory please see one of our appendixes.

3.7 Borel-Weil-Bott theorem of representation of Lie groups

Let G be a compact connected Lie group. A linear representation of G is a homomorphism $\phi : G \to GL(V)$ with V as a vector space. If there is no invariant subspace except 0 and V then the representation is called irreducible. For example, $SU(N)$ acts on the space of homogenous polynomials of N variables of a fixed degree naturally and this gives an irreducible representation of $SU(N)$.

For each representation, the trace of $\phi(g)$, called character or class function, is invariant under conjugation. It is the fundamental invariant of representations.

For a compact connected Lie group G we have a maximal torus $T \subset G$. The conjugacy class of G is the same as T/W, where $W = N(G,T)/T$ is the Weyl group, $N(G,T) = \{g|gT = Tg\}$.

Since T is commutative, the representation of T is trivial. It is the direct sum of one-dimensional representations. So we have $\phi|_T : T \to GL(V), \phi|_T = \phi_1 \oplus ... \oplus \phi_n, \phi_i$ are one-dimensional representations. A one-dimensional representation is just a multiplication by a number. If we take the derivative to the representation we would have a linear functional defined on the sub-Lie algebra corresponding to $T, d\phi_i \in \mathbf{h}^* : \mathbf{h} \to \mathbf{R}^1$. We call $\{d\phi_1, ..., d\phi_n\} \subset \mathbf{h}^*$ the weight system of ϕ. In the case of adjoint representation $\mathrm{Ad} = d\mathrm{ad} : G \to GL(\mathbf{g}), \mathrm{ad} : g \to hgh^{-1}, g, h \in G$, the weight system of $\mathrm{Ad}(G) \otimes \mathbf{C}$ is called the root system, $\Delta(G)$.

Let f be a class function on G, i.e. its value only depends on conjugacy classes of G, then we have Weyl's integral formula:

$$\int_G f(g)dg = \frac{1}{|W|}\int_T f(t)|Q(t)|^2 dt,$$

where we take the Haar measure on G and $|W|$ the number of Weyl chambers,

$$Q(\mathrm{Exp}H) = \sum_{\sigma \in W} \mathrm{sign}(\sigma) e^{2\pi i(\sigma(\delta), H)}, t = \mathrm{Exp}H, H \in \mathbf{h},$$

$\delta = \frac{1}{2}\sum_{\alpha \in \Delta^+(G)} \alpha, \Delta^+$ is the set of positive roots. The $|Q(t)|^2$ factor is obtained from the consideration of Jacobians.

Based on the formula above one can derive the following fundamental Weyl's character formula. Again let $\phi : G \to GL(V)$ be an irreducible representation of G. We then have the highest weight Λ_ϕ whose multiplicity can be shown to be one. Weyl's character formula is then:

$$\chi_\phi = \mathrm{tr}(\mathrm{Exp}H) = \frac{\sum_{\sigma \in W} \mathrm{sign}(\sigma) e^{2\pi i(\sigma(\Lambda_\phi + \delta), H)}}{\sum_{\sigma \in W} \mathrm{sign}(\sigma) e^{2\pi i(\sigma(\delta), H)}},$$

where $H \in \mathbf{h}, \sigma \in W = N(G,T)/T, \delta = \frac{1}{2}\sum_{\alpha \in \Delta^+(G)} \alpha, (\sigma(\delta), H)$ is the natural pairing. In particular, this implies that the highest weight determines all characters and hence the representation.

It is remarkable that all irreducible representations of a compact connected Lie group can be constructed geometrically from the work of Borel-Weil-Bott. We outline it as follows.

Again let $T \subset G$ be a maximal torus. It can be shown that G/T is a natural symplectic variety. G/T is also a complex variety whose complex structure comes from $G_{\mathbf{C}}/B_{\mathbf{C}}, B_{\mathbf{C}}$ the Borel subgroup. One can construct a natural line bundle over G/T whose first Chern class agrees with the natural symplectic form. G acts on G/T by translation and hence on the space of cohomology of sections of the line bundle. This induces a linear representation of G. More precisely, we have the following.

Let $\lambda \in \mathbf{h}^*$ be an integral valued linear functional.

a) If $< \lambda+\delta, \alpha >= 0$, for some $\alpha \in \Delta$, then $H^{0,k}(G/T, \mathbf{C}_\lambda) = 0$, for any k.

b) If $< \lambda+\delta, \alpha >\neq 0$, for all $\alpha \in \Delta$. Let $q = \sharp\{\alpha \in \Delta^+ | < \lambda+\delta, \alpha >< 0\}$. Choose $w \in W$ with $w(\lambda+\delta)$ dominant. Put $\mu = w(\lambda+\delta) - \delta$, then

$$H^{0,k}(G/T, \mathbf{C}) = 0, k \neq q, \mathrm{or} F^\mu, k = q.$$

F^μ is then a finite dimensional irreducible representation of G with the highest weight μ. For more details, see [Knapp-Vogan].

Chapter 4

Chern-Simons-Witten Theory and Three Manifold Invariant

4.1 Representation of mapping class group and three manifold invariant

In quantizing Chern-Simons theory we have constructed a Hilbert space vector bundle $\Gamma^J_{hol}(\mathcal{M}, \mathcal{L}^K) \to$ Teich over the space of complex structures on Σ. We explained that there exists a projective flat connection of the above vector bundle.

The above vector bundle is also a bundle over the moduli space of complex structures, i.e. the quotient space of Teichmuller space module the action of the mapping class group. The mapping class group of Σ is the group of outer automorphisms of $\pi_1(\Sigma)$ or which amounts to be the same, the set of different choices of generators of $\pi_1(\Sigma)$. It acts on the space of complex structures.

With the above projective flat connection over this space, we consider the holonomy of the projective flat connection. This gives a projective representation of the mapping class group. One can use this to define knot invariant and three manifold invariant.

In the case of punctured sphere, the connection is given by the Knizhik-Zamolodchikov equations. Tsuchiya and Kanie [Tsuchiya-Kanie] evaluated the monodromy of Knizhik-Zamolodchikov equations which gives Jones polynomials. Moore and Seiberg [?] used polynomial equations to define the representation of mapping class groups. Kohno [Kohno] used their representation to define topological invariant of three manifold which gives another realization of Witten's invariant of three manifold. The first rigorous realization of Witten's invariant was given by Reshetikhin-Turaev

[Reshetikhin-Turaev] using quantum groups.

4.1.1 *Knizhik-Zamolodchikov equations and conformal blocks*

As we explained in the last chapter, the Hilbert space for a punctured sphere is simply the tensor products, $V_{j_1} \otimes V_{j_2} \otimes ... \otimes V_{j_n}$, of representations. At each puncture P_k we assoiciate one spin j_k representation V_{j_k} of $sl(2, \mathbf{C})$ which is an irreducible representation of dimension $2j+1, 0 \leq j_k \leq \frac{K}{2}, 1 \leq k \leq n$, K is a fixed integer. We know that it is a vector bundle of complex structures over the punctured sphere. In this case it is simply $\bar{\mathbf{C}}^n - \Delta$, where $\Delta = \{(z_1, z_2, ..., z_n)\}$. So we have a trivial vector bundle

$$V_{j_1} \otimes V_{j_2} \otimes ... \otimes V_{j_n} \to \bar{\mathbf{C}}^n - \Delta.$$

The projective flat connection can be explicitly described in this case.

Let $\{I_\mu\}$ be an orthonomal basis of $sl(2, \mathbf{C})$ with respect to the Cartan-Killing form. Let

$$\Omega_{ij} = \Sigma_\mu \pi_i(I_\mu)\pi_j(I_\mu) \in \mathrm{End}(V_{j_1} \otimes V_{j_2} \otimes ... \otimes V_{j_n}),$$

where π_i is the projection from $V_{j_1} \otimes V_{j_2} \otimes ... \otimes V_{j_n}$ to V_{j_i}.

The Knizhik-Zamolodchikov equations are:

$$\frac{\partial \Phi}{\partial z_i} - (\frac{1}{K+2}\Sigma_{j \neq i}\frac{\Omega_{ij}}{z_i - z_j})\Phi = 0;$$

$$\frac{\partial \Phi}{\partial \bar{z}_i} = 0, i = 1, 2, ..., n,$$

where Φ is a section of the Hilbert space vector bundle. The fundamental fact is that the left hand first order differential operators are commutative. Hence it gives a projective flat connection. The commutativity follows from the following two facts about Ω_{ij},

$$[\Omega_{ab}, \Omega_{cd}] = 0, a, b, c, d \text{ distinct},$$

$$[\Omega_{ac}, \Omega_{ab} + \Omega_{bc}] = 0, a, b, c \text{ distinct}.$$

Hence we have an integrable connection

$$\omega = \frac{1}{K+2} \Sigma_{1 \leq i < j \leq n} \Omega_{ij} d \log(z_i - z_j).$$

The solutions of Knizhik-Zamolodchikov equations are vectors of the Hilbert space. They are also horizontal sections of the Hilbert space vector bundle. To find a solution is to evaluate the holonomy of the above connection along a loop. Such solutions are given by iterated path integrals by K. T. Chen [Chen].

We may take advantage of the existence of the flat connection by decomposing the surface into simple pieces. Let $\{\alpha_1, ..., \alpha_{2g-2}\}$ be a set of simple closed curves such that $\Sigma - \{\alpha_1, ..., \alpha_{2g-2}\}$ consisting of pants, i.e. a punctured spheres with three boundary components. We may take the complex structure to be nearly degenerate along the simple closed curves and then the Hilbert space will be the tensor products of those Hilbert spaces of pants which we called conformal blocks.

Let Σ be a pant decomposition. We associate a dual graph γ to it. To each pant we associate a vertex and to each simple closed curve we associate an edge.

Let $K > 0$ be a positive integer. We construct the Hilbert space $Z_K(\gamma)$ as follows. It consists of basis

$$f : \text{edge} \to \{0, 1/2, ..., K/2\} = \mathcal{P}_K,$$

which satisfies the following conditions:

$$|f(c_1) - f(c_2)| \leq f(c_3) \leq f(c_1) + f(c_2),$$

$$f(c_1) + f(c_2) + f(c_3) \in \mathbf{Z},$$

$$f(c_1) + f(c_2) + f(c_3) \leq K.$$

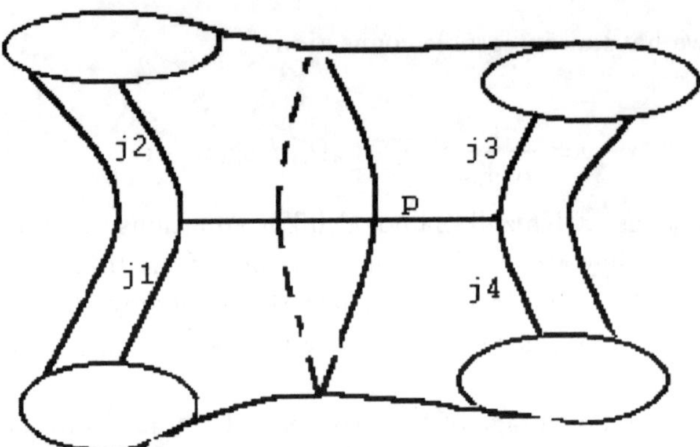

Fig. 4.1 Four punctured sphere

One may think of $f(c)$ as the weight of the associating irreducible representation of the gauge group to a simple close curve on the surface. The first two conditions are Clebsch-Gordon conditions. Let j_1, j_2 be weights of representations V_{j_1}, V_{j_2}, we can decompose the tensor product $V_{j_1} \otimes V_{j_2}$ into irreducible representations $\oplus_j V_j$, then j_1, j_2, j must satisfy the first two conditions.

4.1.2 *Braiding and fusing matrices*

We indicated above that the Hilbert space for general surfaces can be constructed from conformal blocks. The action of mapping class group also localize to those action on simple pieces. In the following we will analyse those elementary objects.

Let us first consider the case of four punctured sphere. Choose a simple closed curve to separate the four punctured sphere into two pieces of three punctured spheres. In graphic representation, Fig. 4.1 illustrates a four punctured sphere and the dual graph of a decomposition of the surface into pants. At each edge we associate a representation of the gauge group. We have two choices of such separating simple closed curves. For each choice we have a Hilbert space. There is a matrix to connect those two Hilbert spaces and we call this a **fusing matrix**, see Fig. 4.2.

Algebraically we have the expression:

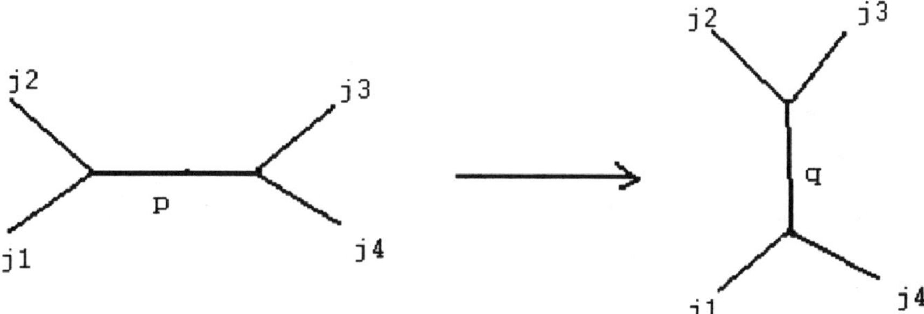

Fig. 4.2 Fusing

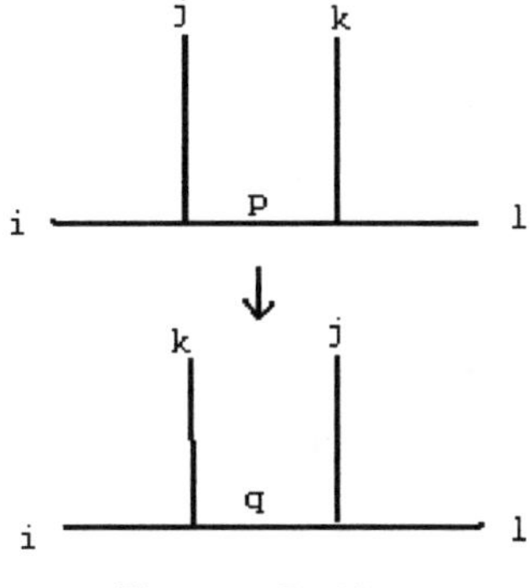

Fig. 4.3 Braiding

$$F : \oplus_r V^i_{jr} \otimes V^r_{kl} \rightarrow \oplus_s V^i_{sl} \otimes V^s_{jk}.$$

We may consider to exchange the two representations j and k of the dual graph. The map between the two Hilbert spaces is called a **braiding matrix**. See Fig. 4.3.

Algebraically it is:

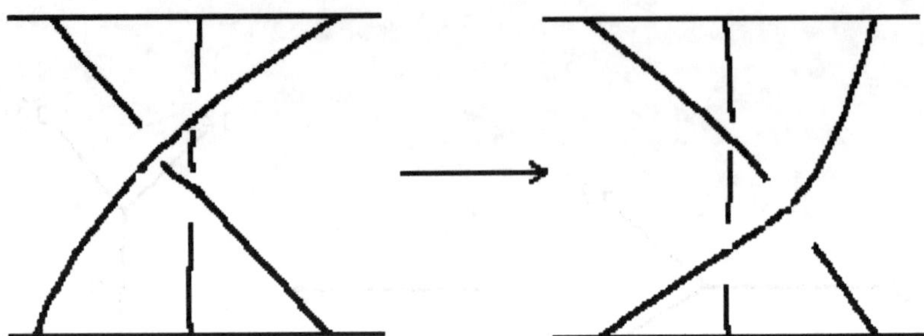

Fig. 4.4 Yang-Baxter for braiding matrices

Fig. 4.5 Yang-Baxter for fusing matrices

$$B : \oplus_p V^i_{jp} \otimes V^p_{kl} \to \oplus_q V^i_{kq} \otimes V^q_{jl}.$$

From the Knizhik-Zamolodchikov equations, we can prove that braiding matrices and fusing matrices satisfies Yang-Baxter properties, see Fig. 4.4, Fig. 4.5.

Braiding matrices and fusing matrices are connected by:

$$B_{ij}(j_2, j_3; j_1, j_4) = (-1)^{j_1+j_4-i-j} \exp(\pi\sqrt{-1}(\Delta_{j_1}+\Delta_{j_4}-\Delta_i-\Delta_j))F_{ij}(j_1, j_3; j_2, j_4).$$

It is true because if we consider two operations $P(x \otimes y) = y \otimes x, C_i^{j_2 j_3}$:

$V_{j_2} \otimes V_{j_3} \to V_i$, then we have $PC_i^{j_2 j_3} = (-1)^{j_2+j_3-i} C_i^{j_2 j_3}$.

We have $\Delta_j = \frac{j(j+1)}{2}$, and it arises as follows. Let ω be the integrable Knizhik-Zamalodchikov connection. It acts naturally on the space $\mathrm{Hom}_{sl(2,\mathbf{C})}$
$(V_{j_1} \otimes ... \otimes V_{j_n}, V_{j_{n+1}})$. Let $\omega_{i_1 ... i_k} = 0$ be the exceptional divisor corresponding to the inverse image of the subspace of $\mathbf{C}^n$ defined by $z_{i_1} = ... = z_{i_k}$. Let $C_{\gamma,f}$ be a vector in the Hilbert space, then we have the residue of ω on the exceptional divisor is:

$$\mathrm{Res}_{\omega_{i_1 ... i_k}=0} C_{\gamma,f} = \Delta(i_1, ..., i_k) C_{\gamma,f},$$

where $\Delta(i_1, ..., i_k) = \Delta_{f(a_{i_1}, ..., i_k)} - \Sigma_{p=1}^k \Delta_{f(a_{i_p})}, \Delta_j = \frac{j(j+1)}{K+2}$.

$\Phi_{\gamma,f} = \Pi \omega_{i_1 ... i_k}^{\Delta(i_1, ..., i_k)} (C_{\gamma,f} + \text{higher order holomorphic terms})$ is a solution of Knizhik-Zamalodchikov equations.

Next we consider a torus. The Hilbert space is given by Verlinde basis $\{v_0, v_{1/2}, ..., v_{k/2}\}$, where v_i is an integrable highest weight representation of the affine Lie algebra $A_1^{(1)}$. Each such representation is determined by an irreducible representation of the underlying Lie algebra. There are two modular transformations:

$$S v_i = \Sigma_j s_{ij} v_j,$$

$$T v_j = \exp 2\pi\sqrt{-1}(\Delta_j - \frac{c}{24}) v_j, 1 \le i, j \le n.$$

S, T are called switching operators. It gives a representation of the modular group $SL(2, \mathbf{Z})$ which is the mapping class group of the torus.

4.1.3 *Projective representation of mapping class group*

With the construction above of representations of mapping class group of simple pieces we can construct representations of mapping class group for general surfaces. This was first realized by Moore and Seiberg in [Moore and Seiberg]. Here we follow expositions of Kohno [Kohno].

Let Σ be a surface of genus g. Let $\{\alpha_i, \delta_i, \beta_i, \epsilon_i\}$ be a set of simple closed curves.

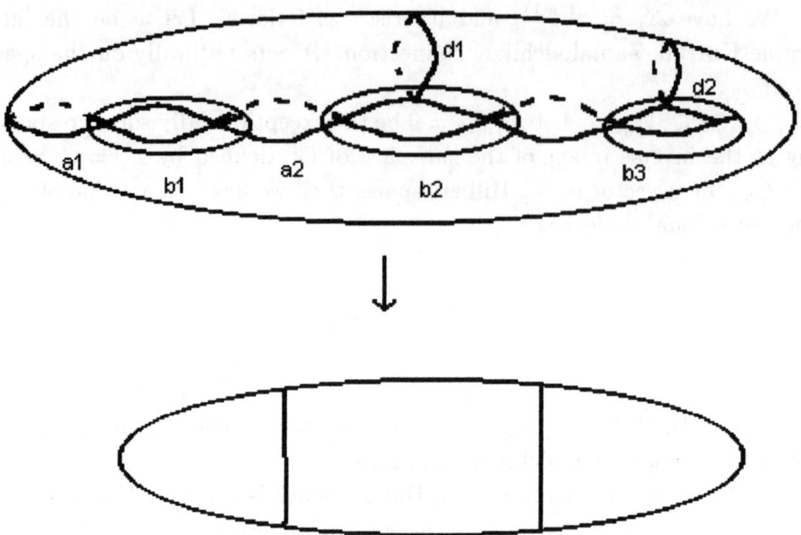

Fig. 4.6 A surface of genus g

We use the same notation to denote Dehn surgery along those curves. It is known that this set of Dehn surgeries generate the full mapping class group. The set $\{\alpha_i, \delta_i, \epsilon_i\}$ are disjoint simple closed curves. The dual graph for this set of simple closed curves is λ. We will construct a projective representation of the mapping class group with respect to this dual graph.

In this dual graph, $\{\alpha_i, \delta_i, \epsilon_i\}$ correspond to edges. The action of mapping class group with respect to λ is very easy, they just change the weight of the corresponding edge. For example, we have

$$\alpha_1 v_i = \exp(-2\pi\sqrt{-1}\Delta_{t_i})v_i.$$

$$\bar{\alpha} = T_a^{-1},$$

$$T_a e_{\gamma(\mu),f} = \exp 2\pi\sqrt{-1}(\Delta_{f(a)} - \frac{c}{24})e_{\gamma(\mu),f}.$$

To describe actions of β_i, we have to use switching operators. β_1, β_g acts

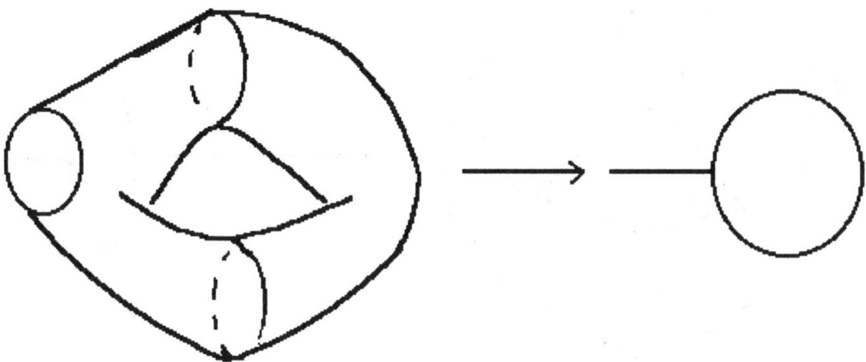

Fig. 4.7 Once punctured torus

on the Hilbert space of once punctured sphere. We have $\bar{\beta_1} = T_{a_1} S_{b_1} T_{a_1}$, where $\bar{\beta_1}$ acts on conformal blocks with dual graph $\Gamma_1, f(b_1) = j_1$.

For $1 \leq k \leq g$, we have

$$\bar{\beta_k} = T_{c_k} S_{b_{k-1} b_k} T_{c_k},$$

where b_{k-1}, b_k are two edges of the conformal blocks Γ_2.

Finally we reach the fundamental result of Moore and Seiberg.

Let Γ_K be the cyclic group generated by $\exp 2\pi\sqrt{-1}(c/24)\mathrm{id} \subset GL(Z_K(\gamma))$, and $c = \frac{3K}{K+2}$, then the above construction of $\bar{\alpha_i}, \bar{\delta_i}, \bar{\beta_i}, \bar{\epsilon_i}$ gives a projective representation of the mapping class group $\rho_K : \mathcal{M} \to GL(\Gamma_K(\gamma))/\Gamma_K$.

Here is an example of the representation of mapping class group.

Let $K = 1, \gamma$ be the dual graph of the following decompositions:

$f(b_1) = f(b_2) = ... f(b_{g-1}) = 0, f(a_k) = f(c_k)$

are admissible weights. So we have the vector space $Z_1(\gamma) = V^{\otimes g}, \dim_{\mathbf{C}} V = 2$. We have:

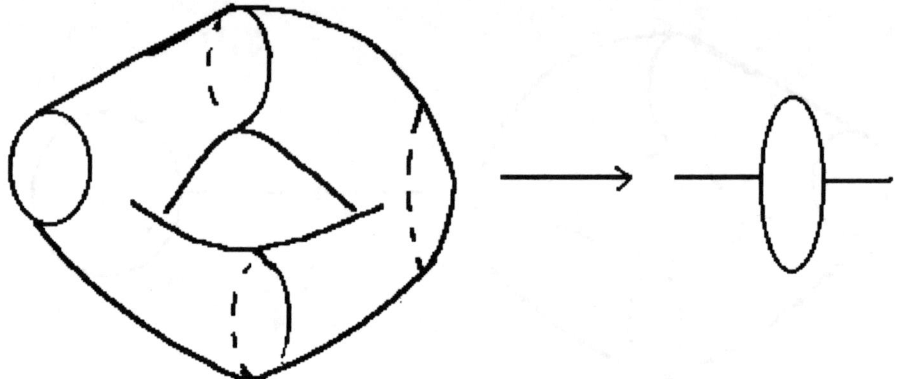

Fig. 4.8 Twice punctured sphere

$$S = \frac{1}{\sqrt{2}} \begin{pmatrix} 1 & 1 \\ 1 & -1 \end{pmatrix}.$$

$$T = \mathrm{diag}(1, i).$$

$$U = STS = \frac{1}{\sqrt{2}} \begin{pmatrix} 1 & i \\ i & 1 \end{pmatrix}$$

$$W = T^{-1} \otimes T^{-1} = \mathrm{diag}(1, -i, -i, 1),$$

$$F \begin{pmatrix} j_2 & j_3 \\ j_1 & j_4 \end{pmatrix} = 1, \text{if one of } j_i = 0.$$

We have the representation of $\rho_1 : \mathcal{M}_g \to GL(V^{\otimes g})/\Omega_g$.

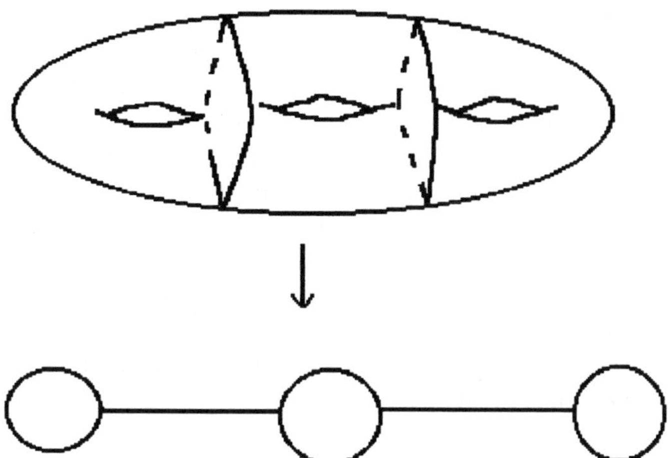

Fig. 4.9 A decomposition of surface and its dual graph

$$\rho_1(\alpha_1) = T_1^{-1} = T^{-1} \otimes \ldots \otimes 1,$$

$$\rho_1(\alpha_2) = w_{12}, \ldots, \rho_1(\alpha_g) = w_{g-1,g},$$

where $w_{k,k+1}$ is the operation of w on $k-$th and $(k+1)-$th components of $V^{\otimes g}$.

$$\rho_1(\delta_2) = T_2^{-1} = 1 \otimes T^{-1} \otimes \ldots \otimes 1,$$

$$\rho_1(\beta_1) = U_1, \ldots, \rho_1(\beta_g) = U_g.$$

4.1.4 *Three-dimensional manifold invariants via Heegard decomposition*

Let M^3 be a three manifold. It admits a Heegard decomposition:

$$M^3 = V_1 \cup_\Sigma V_2,$$

where $\Sigma \subset M^3$ is an embedded surface such that $\partial V_1 = \partial V_2 = \Sigma, V_1 \cap V_2 = \Sigma$.

Taking another point of view, we may take a handlebody V_1, its second copy V_2 and a homeomorphism $h : \partial V_1 \to \partial V_2$. We then construct a three manifold $M^3 = V_1 \cup_h V_2$ by identifying points in ∂V_1 with points in ∂V_2 via h, an element of the mapping class group of the surface ∂V_1.

For any triangulation of M^3, consider its dual and their regular tabular neighborhood of their 1-skeleton, it gives a Heegard decomposition. For S^3, consider the standard sphere S^2, it gives a Heegard decomposition of the three sphere.

From our previous work, we have constructed projective representations of the mapping class group:

$$\rho : \mathcal{M}_g \to GL(Z_K(\gamma))/\Gamma_K.$$

Let $e_{\gamma,0}$ be the vector in $Z_K(\gamma)$ with admissible weight $f : \mathrm{edge}(\gamma) \to \mathcal{P}_K$, such that $f(a) = 0$ for any $a \in \mathrm{edge}(\gamma)$. We have:

$$\rho_K(h)e_{\gamma,0} = \rho_K(h)_{00}e_{\gamma,0} + \Sigma_{f\neq 0}\rho_K(h)_{f,0}e_{\gamma,f}.$$

From our previous work, for two different decompositions of the surface $\gamma(\mu_1), \gamma(\mu_2)$, we have $Z_K(\gamma(\mu_1)) = Z_K(\gamma(\mu_2))$, and the isomorphism send the vector $e_{\gamma(\mu_1),0}$ to $e_{\gamma(\mu_2),0}$. So we see that $\rho_K(h)_{00}$ does not depend on a choice of marking.

There is an important operation on Heegard decomposition. Let $M = V_1 \cup_h V_2$ be a Heegard decomposition of genus g and let $S^3 = D_1 \cup_\tau D_2$ be the standard decomposition of S^3, where D_1 and D_2 denote solid tori. By considering the connected sum of these Heegard decomposition, we obtain a Heegard decomposition of $M\sharp S^3 = M$. We denote by $\tilde{h} : \partial V_1 \sharp \partial D_1 \to \partial V_2 \sharp \partial D_2$ the corresponding attaching homeomorphism. This gives a Heegard decomposition of genus $g+1$. We called this **elementary stabilization**. It is known that Heegard decompositions of a three manifold differ from each other by a number of elementary stabilizations.

It is easy to show that

$$\rho_K(\tilde{\mathrm{h}})_{00} = \sqrt{\frac{2}{K+2}} \sin \frac{\pi}{K+2} \rho_K(h)_{00}.$$

This follows from

$$Z_K(\tilde{\gamma}) = Z_K(\gamma) \otimes Z_K(\Gamma_1, b_g; 0),$$

$$\rho_K(\tilde{h}) e_{\tilde{\gamma},0} = \rho_K(h)_{00} \rho_K(\tau)_{00} e_{\tilde{\gamma},0} + \ldots,$$

$$\rho_K(\tau)_{00} = s_K(0)_{00} = \sqrt{\frac{2}{K+2}} \sin \frac{\pi}{K+2}.$$

From the above it is clear that

$$\phi_K(M^3) = s_K(0)_{00}^{-g} \rho_K(h)_{00}$$

is independent of marking and elementary stabilization. Hence it is a topological invariant. It is a realization of Witten's invariant by Kohno.

4.2 Calculations by topological quantum field theory

4.2.1 *Atiyah's axioms*

M. Atiyah formulated axioms of topological quantum field theory for three dimensional manifold as follows:

To each surface $\Sigma \subset \partial M^3$, one associate a Hilbert space $\mathcal{H}_\Sigma$. To each corbodism $\partial M = \Sigma_1 \cup \Sigma_2$, one associate a morphism $\phi : \mathcal{H}_{\Sigma_1} \to \mathcal{H}_{\Sigma_2}$.

It satisfies the following axiom called naturality. If $\partial M_1^3 = \Sigma_1 \cup \Sigma_2, \partial M_2^3 = \Sigma_2 \cup \Sigma_3, \phi_1 : \mathcal{H}_{\Sigma_1} \to \mathcal{H}_{\Sigma_2}, \phi_2 : \mathcal{H}_{\Sigma_2} \to \mathcal{H}_{\Sigma_3}$, then for $M^3 = M_1^3 \cup_{\Sigma_2} M_2^3$, and $\partial M^3 = \Sigma_1 \cup \Sigma_3$, we have $\phi : \mathcal{H}_{\Sigma_1} \to \mathcal{H}_{\Sigma_3}$, and $\phi = \phi_2 \phi_1$.

This naturality reflects a crucial property of probability. The probability from an initial state to a final state is equal to the sum of product of probability from the initial state to an immediate state and the probability of the immediate state to the final state. Feynman's path integral reflects just this property.

$$< \phi_1|e^{-tH}|\phi_2 >= \int_{\phi(0)\in\phi_1,\phi(1)\in\phi_2} \mathcal{D}\phi e^{iL(\phi)}.$$

H is the Hamiltonian, L is the Lagrangian. A key observation for Chern-Simons theory is that the Hamiltonian is zero, so we only need to consider the inner product in $\mathcal{H}_\Sigma$. We will introduce some applications assuming that Chern-Simons topological quantum field theory exists. In the next chapter we will explain how Chern-Simons perturbation series makes sense with respect to some regulizations.

4.2.2 *An example: connected sum*

Let M_1, M_2 be two three manifold and $M = M_1\sharp M_2$ their connected sum. Applying the axiom of TQFT(Topological Quantum Field Theory) to this case we have $Z(M)Z(S^3) = Z(M_1)Z(M_2)$. Or

$$\frac{Z(M)}{Z(S^3)} = \frac{Z(M_1)}{Z(S^3)}\frac{Z(M_2)}{Z(S^3)}.$$

Remarks : 1) If m_1, m_2 are two arbitary manifolds without knots in it, then the above formula implies multiplicity of Reidemeister and Ray-Singer torsion under connected sums.

2) If we consider Chern-Simons theory with the non-compact gauge group $SL(2,\mathbf{R})$, a leading term in $Z(M)$ gives $e^{i\text{vol}(M)}$. Hence we have the formula $\text{vol}(M) = \text{vol}(M_1) + \text{vol}(M_2)$.

4.2.3 *Jones polynomials*

Let $L = \cup_{C_i} \subset S^3$ be a link, we define the partition function

$$Z(L) = \int \mathcal{D}A e^{\frac{ik}{2\pi}\int_{S^3}\text{Tr}(A\wedge dA+\frac{2}{3}A\wedge A\wedge A)} \prod_i Tr_{R_i} P\exp\int_{C_i} A.$$

Let $G = SU(N)$ and take the representation to be the defining representation of G. We call $Z(L)$ the Jones polynomial.

The most remarkable property for Jones polynomials is the skein relation. We will establish this relation. For a crossing of the knot we imagine inserting a small sphere around the crossing. The sphere would intersect

  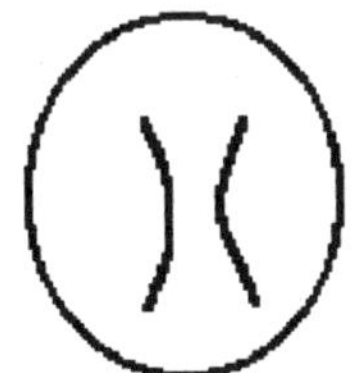

Fig. 4.10 Skein relation

with the knot at four points. The manifold itself can be representated as $M = M_L \cup M_R$, with M_R the three ball and M_L the rest of the three ball. M_R and M_L share the same boundary $\Sigma = S^2 - \{z_1, z_2, z_3, z_4\}$. Consider the Feymann path integral on M_R, M_L, we have vectors $\psi \in \mathcal{H}_R, \chi \in \mathcal{H}_L$ respectively. Notice that $\mathcal{H}_L = \mathcal{H}_R^*$. The partition function is then $Z(L) = (\chi, \psi)$.

If we replace M_R with two other different ways of strings, we would get two other vectors ψ_1, ψ_2.

Since $\dim \mathcal{H}_\Sigma = 2$, we have the three vectors ψ, ψ_1, ψ_2 which satisfy a linear relation: $\alpha\psi + \beta\psi_1 + \gamma\psi_2 = 0$. If we multiply the above by χ, we then have $\alpha Z(L) + \beta Z(L_1) + \gamma Z(L_2) = 0$. This is called the skein relation. It is known that this relation is enough to compute knot invariant. Because we can project a knot into a plane and apply this relation, the number of crossings is decreased eventually.

The novelty here is that we have achieved a three dimensional interpretation of skein relation without projecting a knot into a plane. Invariants defined this way are intrinsic and natural.

4.2.4 *Surgery*

The construction of surgery is the following. Consider an embedded torus Σ in a manifold M, the manifold can be represented as a union of the solid torus and the rest of the manifold with common boundary Σ. We then consider an element $\rho \in SL(2, \mathbf{Z})$, and we glue the two manifolds with boundary Σ after we apply for ρ to the torus, then we obtain a new manifold. This construction is called **surgery**.

In Chapter 3, we calculated the dimension of the Hilbert space for a torus. Consider a simple closed curve on the torus. After collapsing the curve we have twice punctured sphere. The Hilbert space is generated by irreducible representations, or level k integrable highest weight representations of the loop group.

The mapping class group of the torus $SL(2, \mathbf{Z})$ acts on the Hilbert space, $K = \rho^* : \mathcal{H} \hookleftarrow$. We can express it as follows:

$$K.v_i = \sum_j K_i^j v_j,$$

where v_i is a set of basis of $\mathcal{H}$. This is good enough to calculate the change of partition function with respect to surgery.

For the loop group, we define the character of an integrable highest weight representation α as:

$$\chi_\alpha(\tau) = \sum_\lambda \text{multi}_\alpha(\lambda) e^{\dim \lambda \tau},$$

where λ is a grading of the representation. Let us consider $S = \begin{pmatrix} 0 & -1 \\ 1 & 0 \end{pmatrix}$.

Its action on the characters is:

$$\chi_\alpha(-\frac{1}{\tau}) = \sum_\beta S_{\alpha\beta} \chi_\beta(\tau).$$

For the group $G = SU(2)$, we have

$$S_{ij} = \sqrt{\frac{2}{k+2}} \sin \frac{\pi(i+1)(j+1)}{k+2}.$$

Here are some interesting applications:
1) Partition function of the manifold $X \times S^1$ is

$$Z(X \times S^1) = \text{Tr}_{\mathcal{H}_X}(1) = \dim \mathcal{H}_X.$$

For example, we have $Z(S^2 \times S^1) = 1$.

2) Consider a circle along the S^1 direction of the manifold $S^2 \times S^1$ and applying surgery S to this, one gets S^3. By our surgery formula above, we have:

$$Z(S^3) = \sum_j S_0^j Z(S^2 \times S^1; R_j) = S_{0,0} = \sqrt{\frac{2}{k+2}} \sin \frac{\pi}{k+2}.$$

3) Consider three punctured sphere $X = S^2 - \{z_1, z_2, z_3\}.X \times S^1$ can be considered as $S^2 \times S^1$ with three unknotted Wilson lines. To each line, we associate a representation R_i, R_j, R_k, then we have

$$Z(X \times S^1) = \dim \mathcal{H}_X = N_{ijk}.$$

4.2.5 *Verlinde's conjecture and its proof*

Consider now two unknotted and unlinked knot R_j, R_k in S^3. Consider another knot R_i which linked R_j, R_k. We call this configuration $L(R_i, R_j, R_k)$. By applying surgery formula to the similar configuration in $S^2 \times S^1$, we obtain

$$Z(S^3; L(R_i, R_j, R_k)) = \sum_m S_i^m Z(S^2 \times S^1; R_m, R_j, R_k) = \sum_m S_i^m N_{mjk}.$$

Consider cutting the link at R_i into two pieces and applying the multiplicative formula, we have

$$Z(S^3; L(R_i, R_j, R_k)) = \frac{S_{ij} S_{ik}}{S_{0i}}.$$

So we have $\frac{S_{ij} S_{ik}}{S_{0,i}} = \sum_m S_i^m N_{mjk}$. Let $\{v_i\} \in \mathcal{H}_{T^2}$ be a basis. Then we have

$$v_i v_j = \sum_k N_{ij}^k v_k.$$

If we let $w_i = S_{0,i} \sum_m S_i^m v_m$, then we have

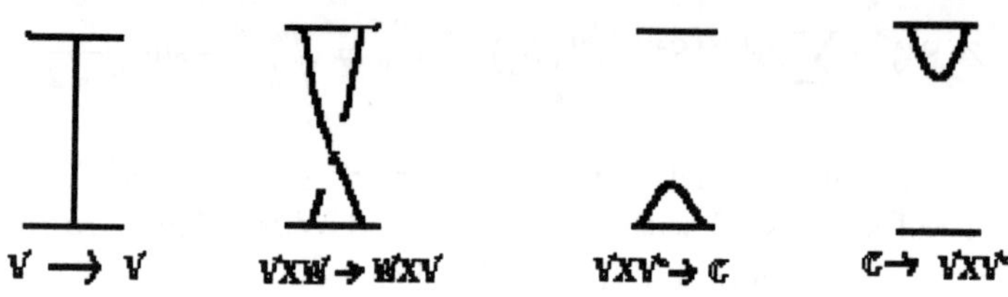

Fig. 4.11 Algebras to tangles

$$w_i w_j = \delta_{ij} w_j.$$

So after we apply a unitary representation S to $\{v_i\}$, we get a new basis. The Verlinde algebra is diagonized with respect to this new basis.

4.3 A brief survey on quantum group method

4.3.1 *Algebraic representation of knot*

To define knot invariant is to a large extent finding a good algebraic representation of knot. One of the ways is to use tangles. We refer to Sawin's paper for more details [Sawin]. A tangle is the image of a smooth embedding of a union of circles and intervals into the cylinder $D \times I$, where D is the unit disk in $\mathbf{C}$ and $I = [0, 1]$.

We want to associate algebras to tangles. To each point of intersection in the line we associate a vector space V. To two points of intersection of the line we associate tensor product $V \otimes W$ of the two vector spaces V and W. When we move the line the vector spaces changes so we will associate operators to the vector spaces.

By a theorem of Markov, a knot can be representated as a braid glued at their two ends. A braid is an element of $\pi_1(\bar{\mathbf{C}} - \{z_1, ..., z_k\})$. If we want to define knot invariant via braid representation it has to be invariant under the following operation.

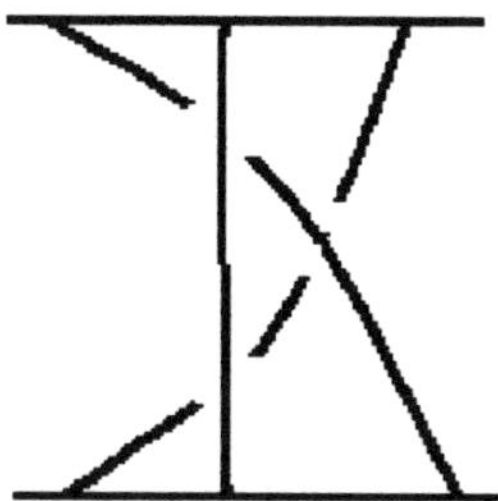
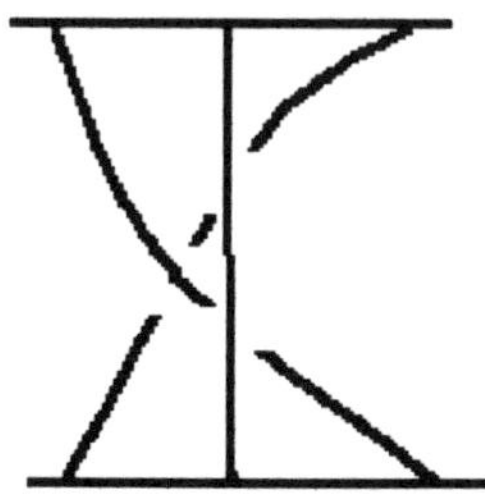

Fig. 4.12 Braids

Define

$$R = \sum_i a_i \otimes b_i,$$

to be the matrix of $V \otimes V \to V \otimes V$. To respect the above move, we have the Yang-Baxter relation:

$$R_{12}R_{13}R_{23} = R_{23}R_{13}R_{12},$$

where

$$R_{13} = \sum_i a_i \otimes 1 \otimes b_i, R_{23} = \sum_i 1 \otimes a_i \otimes b_i, R_{12} = \sum_i a_i \otimes b_i \otimes 1.$$

A real problem arises for the above algebraic representations. That is: given an associative algebra $\mathcal{B}$, and two representations $\rho_1, \rho_2 : \mathcal{B} \to \mathrm{End}(V), \mathrm{End}(V')$, is there a natural way to define a tensor product of ρ_1, ρ_2, i.e. $\rho : \mathcal{B} \to \mathrm{End}(V \otimes V')$?

A naive definition is $\rho = \rho_1 \otimes 1 + 1 \otimes \rho_2$. But it is only a representation as a group, not as an algebra. To cure this problem, we introduce the notion of Hopf algebra.

Definition (Hopf algebra): A Hopf algebra $\mathcal{B}$ is an associative algebra with co-product $\Delta : \mathcal{B} \to \mathcal{B} \otimes \mathcal{B}$, co-unit $\epsilon : \mathcal{B} \to \mathbf{C}$ and an anti-pole $\gamma : \mathcal{B} \to \mathcal{B}$. It satisfies the following:

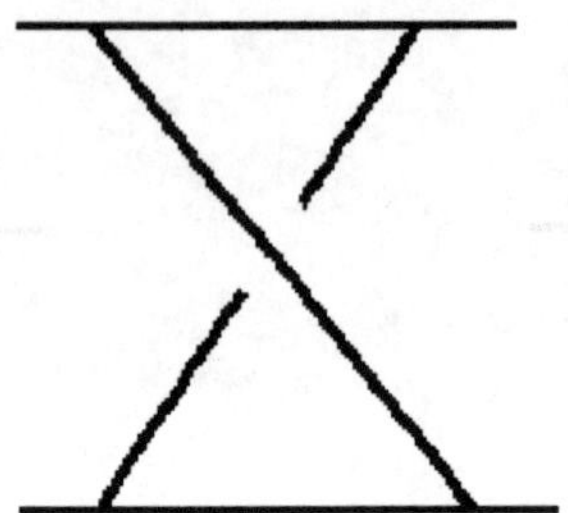

Fig. 4.13 Tangles

$$(\Delta \otimes 1)\Delta = (1 \otimes \Delta)\Delta,$$

$$(\epsilon \otimes 1)\Delta = 1 = (1 \otimes \epsilon)\Delta,$$

$$(id \otimes \gamma)\Delta = \epsilon = (\gamma \otimes id)\Delta.$$

For a Hopf algebra $\mathcal{B}$, and two representations ρ_1, ρ_2 of $\mathcal{B}$, then $\rho = (\rho_1 \otimes \rho_2)\Delta : \mathcal{B} \to \mathrm{End}(V \otimes V')$ is a representation of $\mathcal{B}$.

Here are two examples of Hopf algebra.

1) Let G be a finite group, $\mathbf{C}G$ a group algebra. We define $\Delta(g) = g \otimes g, s(g) = g^{-1}, \epsilon(g) = 1$.

2) Let $\mathcal{U}(\mathrm{g})$ be a universal enveloping algebra. We define $\Delta(x) = x \otimes 1 + 1 \otimes x, \epsilon(x) = 0, \gamma(x) = -x$.

A Hopf algebra staisfying Yang-Baxter relation is called a quasi-triangular Hopf algebra. We define it as follows.

For the tangle in Fig. 4.13, we define $\sigma_{AA'} : V \otimes W \to W \otimes V, \Delta'(a) = \sigma_{AA'}(\Delta(a))$. Then a quasi-triangular Hopf algebra is a Hopf algebra which also satisfy the following:

$$\Delta'(a) = R\Delta(a)R^{-1},$$

$$(\Delta \otimes 1)R = R_{13}R_{23}, (1 \otimes \Delta)R = R_{13}R_{12}.$$

It is just another expression of the Yang-Baxter relation.

For any quasi-triangular Hopf algebra, one can define knot invariant associated to the algebra by taking the trace of the matrix generated by the braid element on the Hilbert vector space.

4.3.2 *Hopf algebra and quantum groups*

Let G be any Lie group, g its Lie algebra and $\mathcal{U}(\mathrm{g})$ be its universal enveloping algebra, i.e. all left invariant differential operators. A quantum group is a deformation of the universal enveloping algebra. Let us illustrate it by an example of $SU(2)$.

The universal enveloping algebra of $SU(2)$ is generated by $\{x, y, h\}$, with relations:

$$[h, x] = 2x, [h, y] = -2y, [x, y] = h.$$

The corresponding Hopf algebra is given by:

$$\Delta(a) = a \otimes 1 + 1 \otimes a, \epsilon(a) = 0, s(a) = -a, a \in \{x, y, h\}.$$

Now we consider deformations of $\mathcal{U}(\mathrm{g})$, denoted as $\mathcal{U}_s(\mathrm{g})$. The relation is modified as:

$$[h, x] = 2x, [h, y] = -2y, [x, y] = \frac{(s^{2h} - s^{-2h})}{s^2 - s^{-2}}.$$

The corresponding Hopf algebra is defined as:

$$\Delta(x) = x \otimes s^h + s^{-h} \otimes x, \Delta(y) = y \otimes s^h + s^{-h} \otimes y, s(x) = -s^2 x, s(y) = -s^2 y.$$

The most interesting ones are those s at the root of unity, $s = \exp(\frac{\pi i}{N})$. It gives modular Hopf algebra.

Let $[k] = \frac{s^k - s^{-k}}{s - s^{-1}}$. If $\{e_0, e_1, ..., e_{N-1}\}$ gives an $N-$ dimensional representation of $sl(2, \mathbf{C})$, then we have N-dimensional representation of $\mathcal{U}_q(sl(2, \mathbf{C}))$ by

$$X(e_i) = [i + 1]e_{i+1}, Y(e_i) = [i]e_{i-1}, K(e_i) = s^{i-(N-1)/2} e_i,$$

where X, Y, K generated $sl(2, \mathbf{C})$ and is deformed into

$$KX = sXK, KY = s^{-1}YK, XY - YX = \frac{K^2 - K^{-2}}{s - s^{-1}}.$$

Quantum groups can be used to construct quasi-triangular Hopf algebras. One can compute the action of R on $V \otimes W$, the composing flip map gives a matrix [Kauffman]:

$$\begin{pmatrix} s & 0 & 0 & 0 \\ 0 & 0 & s^{-1} & 0 \\ 0 & s^{-1} & s - s^{-3} & 0 \\ 0 & 0 & 0 & s \end{pmatrix}.$$

4.3.3 *Chern-Simons theory and quantum groups*

Why does quantum group exist? Witten showed that its existence follows from the existence of Chern-Simons topological quantum field theory. We give a brief introduction here, for more details please see Witten's paper "Gauge theories, vertex models, and quantum groups", **Nuclear Physics** B330 (1990) 285-346.

Let us consider the group $G = SU(2)$. In $SU(2)$ Chern-Simons gauge theory, the physical Hilbert space $\mathcal{H}_{U,A,A,U+1}$ of S^2 with four charges in representations U, A, A and $U + 1$ is two-dimensional.

Proof: We decompose $U \times A = \oplus R_i, R_i = U - 1, U$ and $(U + 1), A \otimes (U + 1) = \oplus \bar{R}_i, \bar{R}_i = U, (U + 1)$ and $(U + 2).U, (U + 1)$ of R_i concide with that of $\bar{R}_i$. This implies that $\dim \mathcal{H}_{U,A,A,(U+1)} = 2$.

Since it is two dimensional, any three vectors obey a linear relation. This is the key to construct an algebra $\mathcal{A}$ over $\mathbf{C}$, generated by symbols T_-, T_0 and T_+ with the following relations. It can be considered as quantum deformation of the $SU(2)$ Lie algebra.

$$T_-T_0 + uT_0T_- + u'T_- = 0,$$

$$T_-T_+ + vT_+T_- + v'T_0^2 + v''T_0 = 0,$$

$$T_+T_0 + wT_0T_+ + w^{'}T_+ = 0.$$

The parameter q in the quantum group is identified as $q = \exp(\frac{2\pi i}{k+2})$, where $\frac{1}{k}$ is the coupling constant of Chern-Simons theory. Among other things, Witten established that:

1) For every non-negative integer of half-integer j, the spin-j representation of $SU(2)$ deform to a representation of the algebra $\mathcal{A}$.

2) There is a natural tensor product of representation of $\mathcal{A}$, and there is an "R-matrix" relating tensor products $j_1 \otimes j_2$ and $j_2 \otimes j_1$.

$$[illegible] = 0.$$

The parameter q in the quantum group is defined as $q = \exp[illegible]$, where k is the coupling constant of Chern-Simons theory. Among other things, Witten established that:

1) For every non-negative integer ℓ or half-integer, the spin-ℓ representation of $SU(2)$ determines [illegible] representation of the algebra [illegible].

2) There is a natural [illegible]

[illegible]

Chapter 5

Renormalized Perturbation Series of Chern-Simons-Witten Theory

5.1 Path integral and morphism of Hilbert spaces

In our previous work of constructing topological quantum field theory, we associated a Hilbert space, $\mathcal{H}_\Sigma$, to each boundary component of a three manifold M^3. Let $\partial M_1^3 = \Sigma_1 \cup \Sigma_2$. We also need to construct a morphism:

$$\phi_{X_1} : \mathcal{H}_{\Sigma_1} \to \mathcal{H}_{\Sigma_2}.$$

According to Atiyah, such morphisms should satisfy the composition law,

$$\phi_X = \phi_{X_2}\phi_{X_1},$$

where $\phi_{X_2} : \mathcal{H}_{\Sigma_2} \to \mathcal{H}_{\Sigma_3}$ is a morphism associated with a manifold M_2^3 with boundary $\Sigma_2 \cup \Sigma_3$ and $\phi_X : \mathcal{H}_{\Sigma_1} \to \mathcal{H}_{\Sigma_3}$ is the morphism associated with the manifold M^3 glued from M_1^3 and M_2^3 at their common boundary Σ_2.

We will construct such morphisms through path integrals.

5.1.1 *One-dimensional quantum field theory*

In this case we have the following axioms:

1) For every point we associate a Hilbert space:

$$. \to \mathcal{H}$$

2) For every one-dimensional manifold with Riemannian structure we associate a morphism:

$$\phi_{I,t} : \mathcal{H} \to \mathcal{H}.$$

The morphism satisfies the composition law:

$$\phi_{I,t_1+t_2} = \phi_{I,t_1}\phi_{I,t_2}.$$

So there exists an operator H on $\mathcal{H}$, such that

$$\phi_{I,t} = e^{-tH},$$

H is called the Hamiltonian.

If the one-dimensional manifold is a circle then we have the morphism as:

$$\phi_{S^1,t} : \mathcal{H}_\phi \to \mathcal{H}_\phi.$$

In this case $\mathcal{H}_\phi = \mathbf{C}$. From the axioms we have that:

$$\phi_{S^1,t} = \mathrm{Tr} e^{-tH}.$$

5.1.2 *Schroedinger operator*

We now consider a special case. Consider the space $\mathbf{R}^{2n} = T^*(\mathbf{R}^n)$ with the standard symplectic form $\omega = \Sigma dp^i \wedge dx^i$. The Hilbert space is $\mathcal{H} = L^2(\mathbf{R}^n)$. We have operator representations x_i by multiplication, and p_i by $-i\frac{d}{dx_i}$.

$$H = \frac{1}{2}\Sigma p_i^2 + V(x_i) = -\frac{1}{2}\Delta_i^2 + V(x_i).$$

This can be generalized to any cotangent bundle of a manifold M. The Hamiltonian operator would be $-\Delta^2 + V$ on M. One-dimensional quantum field theory is essentially the theory of elliptic operators on manifolds. Now we want to construct morphisms via Feynman path integral.

To understand e^{-tH} is to understand its kernel. Given $\psi(x)$, we define the kernel as follows:

$$(e^{-tH}\psi)(x^{''}) = \int dx^{'} K(x^{''}, x^{'}; t)\psi(x^{'}).$$

If we use distributive "wave" functions, e.g. $|x^{'} >= \delta(x - x^{'})$, then

$$K(x^{''}, x^{'}, ; t) = (x^{''}, e^{-tH} x^{'}).$$

Here are more identities:

$$(g, e^{-tH} f) = \int dydx g^{*}(x) K(x, y; t) f(y),$$

$$|x^{'} >= \delta(x - x^{'}), |x^{''} >= \delta(x - x^{''}),$$

$$V(x)|x^{'} >= V(x)\delta(x - x^{'}) = V(x^{'})\delta(x - x^{'}),$$

$$\Sigma p_i^2 e^{ip^{'}x} = p^{'2} e^{ip^{'}x}, p_i = -i\frac{\partial}{\partial x_i}.$$

Here $|p^{'} >$ is the vector given by $\psi_{p^{'}}(x) = \frac{e^{ip^{'}x}}{\sqrt{2\pi}}$.

$$< p^{'}|H|x^{'} >= (p^{'}|\frac{p^2}{2} + V(x)|x^{'})$$

Here $e^{-ip^{'}x^{'}} = (e^{ip^{'}x}, \delta(x - x^{'})), (p^{'}|H|x^{'}) = H(p^{'}, x^{'})\frac{e^{-ip^{'}x^{'}}}{\sqrt{2\pi}}$. If t is small,

$$(p^{'}|e^{-tH}|x^{'}) = (p^{'}|(1-tH)|x^{'}) = (1-tH(p^{'}, x^{'}))\frac{e^{-ip^{'}x^{'}}}{\sqrt{2\pi}} = e^{-tH(p^{'},x^{'})}\frac{e^{-ip^{'}x^{'}}}{\sqrt{2\pi}}.$$

According to Fourier,

$$(x^{''}|e^{-tH}|x^{'}) = \int dp_1...dp_n (x^{''}, p^{'})(p^{'}, e^{-tH} x^{'}) = \int dp_1...dp_n \frac{e^{ip^{'}x^{''}}}{\sqrt{2\pi}} e^{-ip^{'}x^{'}} e^{-tH(p^{'},x^{'})}$$

up to error of t^2. If t is not small, let $\Delta t = \frac{t}{N}, e^{-tH} = (e^{-\Delta t H})^N$, then let $N \to \infty$.

Using it repeatedly,

$$(x'''|e^{-(t_1+t_2)H}|x') = \int dx''(x'''|e^{-t_2 H}|x'')(x''|e^{-t_1 H}|x').$$

Then,

$$(x''|e^{-tH}|x')$$

$$= \int dx_{N-1}dx_{N-2}..dx_1(x'', e^{-\Delta t H}x_{N-1})(x_{N-1}, e^{-\Delta t H}x_{N-2})...(x_1, e^{-\Delta t H}x')$$

$$= \int dP_N dx_{N-1}dp_{N-1}...dx_1 dp_1 e^{i\Sigma_{l=1}^N p_l(x_l - x_{l-1})}e^{-\Delta t \Sigma_{l=1}^N H(p_l, p_{l-1})}.$$

Note that:

$$\int dp_l e^{ip_l(x_l - x_{l-1})}e^{-\Delta t \frac{p_l^2}{2}} = e^{-\frac{1}{2}\Delta t (x_l - x_{l-1})^2 (\frac{1}{\Delta t})},$$

or $\int dp e^{-\frac{1}{2}tp^2 + ipx} = \sqrt{\frac{\pi}{t}}e^{-\frac{1}{2}tx^2}$. So

$$(x'', e^{-tH}x') = \int dx_{N-1}...dx_1 e^{-\Delta t \Sigma_{l=1}^N \frac{(x_l - x_{l-1})^2}{(\Delta t)^2}}e^{-\Delta t \Sigma_l V(x_l)}.$$

According to Feynman, if we take $N \to \infty$, the integral becomes an integral over the space of paths

$$\mathcal{N}_t(x'', x') = \{ \text{ maps } x : [0,1] \to \mathbf{R}^n, \text{ s.t. } x(0) = x', x(t) = x''\}.$$

$$(x'', e^{-tH}x') = \int_{\mathcal{N}_t} \exp(-\int_0^t dt'(\frac{1}{2}(\frac{dx}{dt})^2 + V(x))).$$

We can also consider the space:

$$\mathcal{L}_t(x'', x') = \{ \text{ maps } (x,p) : [0,1] \to \mathbf{R}^{2n}, \text{ s.t. } x(0) = x', x(t) = x'' \}.$$

The integral over $\mathcal{L}_t(x'', x')$ gives phase space version:

$$K(x'', x'; t) = \int_{\mathcal{L}_t(x'', x')} e^{-\int \Sigma p_i dx_i} e^{-\int dtH}.$$

$$= \int \mathcal{D}p\mathcal{D}x e^{\int (ip\frac{dx}{dt} - \frac{1}{2}(p^2+x^2))dt}.$$

5.1.3 *Spectrum and determinant*

Let us consider the space $\mathbf{R}^2 = T^*(\mathbf{R}^1), V = \frac{1}{2}x^2, H = \frac{1}{2}(p^2 + x^2)$. We knew its spectrum to be

$$H\psi_n = (n + \frac{1}{2})\psi_n, \psi_n = (p + ix)^n e^{-\frac{x^2}{2}}, n = 0, 1, 2, ...$$

$$\text{Tr} e^{-Ht} = \Sigma_{n=0}^{\infty} e^{-t(n+\frac{1}{2})} = \frac{1}{e^{\frac{t}{2}} - e^{-\frac{t}{2}}} = \frac{1}{2\sinh\frac{t}{2}} = \frac{1}{t\Pi_{n=1,2,...}(1 + \frac{t^2}{(2\pi n)^2})},$$

$$(x''|e^{-Ht}|x') = \Sigma_n \psi_n^*(x'')\psi_n(x')e^{-(n+\frac{1}{2})t}.$$

For $t = \frac{i\pi}{2}$, we have

$$e^{-\frac{i\pi H}{2}} = \text{Fourier trans. } e^{\frac{i\pi}{4}},$$

$$(x''|e^{-\frac{i\pi H}{2}}|x') = e^{ix''x'} e^{\frac{i\pi}{4}}.$$

We want to calculate this integral. We knew the following formulae of Gaussian integral:

$$\int \frac{du_1...du_n}{(2\pi)^{\frac{n}{2}}} e^{-\frac{1}{2}\Sigma\Lambda_{ij}u_iu_j} = \frac{1}{\sqrt{\det \Lambda}}.$$

We consider operator Λ on pairs $(p(t), x(t))$,

$$\frac{1}{2}(p^2 + x^2) - ip\frac{dx}{dt} = (p, x)\Lambda(p, x)^{tr}.$$

Here, $\Lambda = \begin{pmatrix} 1 & -i\frac{d}{dt} \\ i\frac{d}{dt} & 1 \end{pmatrix}$. Λ has two-dimensional invariant subspaces $e^{\frac{2\pi int}{T}}(p_n, x_n)^{tr}$. In such a subspace,

$$\Lambda = \begin{pmatrix} 1 & \frac{2\pi n}{T} \\ -\frac{2\pi n}{T} & 1 \end{pmatrix}.$$

So formally, $\det \Lambda = \Pi_{n \in \mathbf{Z}} \det \Lambda_n$,

$$\det \Lambda = \Pi_{n \in \mathbf{Z}}(1 + (\frac{2\pi n}{T})^2).$$

$$\frac{1}{\sqrt{\det \Lambda}} = \frac{1}{\sqrt{\Pi_{n \in \mathbf{Z}}(1 + (\frac{2\pi n}{T})^2)}} = \frac{1}{\Pi_{n=1}^{\infty}(1 + (\frac{2\pi n}{T})^2)}$$

After Ray and Singer, given an operator $\mathcal{O} > 0$ and $\mathcal{O}\psi_n = \lambda_n \psi_n$, we define $\zeta(s) = \Sigma \lambda_n^{-s}$, formally $\zeta'(s) = \Sigma_i \lambda_i^s \log \lambda_i$, so

$$\det \mathcal{O} = e^{-\zeta'(0)}.$$

Given $r \in \mathbf{R}^+$, define $\mathcal{O}_r = r\mathcal{O}, \zeta_{r\mathcal{O}}(s) = \Sigma(r\lambda_i)^{-s} = r^{-s}\zeta_{\mathcal{O}}(s)$, then

$$\det(r\mathcal{O})) = \exp(-\zeta'_{r\mathcal{O}}(0)) = \exp(-\frac{d}{ds}(r^{-s}\zeta_{\mathcal{O}}(s)))|_{s=0} = r^{\zeta_{\mathcal{O}}(0)} \det \mathcal{O}.$$

If $\mathcal{O}$ is a finite dimensional matrix on V, then $\zeta_{\mathcal{O}}(0) = \dim V, \det(r\mathcal{O}) = r^{\dim} \det(\mathcal{O})$.

Let $\mathcal{O}$ be an operator with $\det \mathcal{O} = \Pi_n(1 + (\frac{2\pi n}{T})^2)$, and consider $\mathcal{O} = (\frac{2\pi}{T})^2\mathcal{O}'$, then

$$\det \mathcal{O}' = \Pi_n(n^2 + (\frac{T}{2\pi})^2).$$

And in ζ function regulation,

$$\det \mathcal{O} = (\frac{2\pi}{T})^{2\zeta_{\mathcal{O}}(0)} \det \mathcal{O}'.$$

If $\Pi\frac{\lambda_n}{\lambda'_n}$ converges, then in ζ function,

$$\Pi\lambda_n = \Pi\lambda'_n \Pi(\frac{\lambda_n}{\lambda'_n}),$$

$$\Pi_n(n^2 + (\frac{T}{2\pi})^2) = \Pi_n n^2 \Pi_n(1 + (\frac{T}{2\pi n})^2).$$

And the operator $\mathcal{O}''$ with eigenvalues n^2 for $n = 1, 2, ...$ has the ζ function:

$$\zeta_{\mathcal{O}''(s)} = \Sigma n^{-2s} = \zeta_{\text{Riemann}}(2s),$$

$$\det \mathcal{O}'' = \exp(-2\zeta'_{\text{Riemann}}(0)).$$

5.2 Asymptotic expansion and Feynman diagrams

In this section we illustrate calculations of asymptotic expansions for some integrals over a finite dimensional space. We show that it can be described very well by Feynman diagrams.

5.2.1 *Asymptotic expansion of integrals, finite dimensional case*

Let X be a finite dimensional manifold equipped with a measure $d\mu$. Let f be a smooth function on X. We want to evaluate asymptotic expansion of integral $I(t)$ in the following for $t \to \infty$,

$$I(t) = \int_X d\mu e^{itf}.$$

An easy case is $X = \mathbf{R}^n, d\mu = \frac{dx_1 dx_2 \ldots dx_n}{(2\pi)^{\frac{n}{2}}}$ and $f = \frac{1}{2}\Sigma\lambda_{ij}x_i x_j = \frac{1}{2}\Sigma\lambda_i x_i^2$. We have

$$\lim_{\epsilon\to 0}\int \frac{dx}{\sqrt{2\pi}} e^{i\lambda x^2} e^{-\epsilon x^2} = e^{\frac{i\pi}{2}} \frac{\text{sign }\lambda}{\sqrt{|\lambda|}}.$$

So,

$$I(t) = \int \frac{dx_1 dx_2 ... dx_n}{(2\pi)^{\frac{n}{2}}} e^{it\Sigma\frac{1}{2}\lambda_{ij}x_i x_j}$$

$$= \frac{1}{t^{\frac{n}{2}}} e^{\frac{i\pi}{2}} \frac{\Sigma_i \text{ sign}\lambda_i}{\sqrt{|\det\Lambda|}}$$

$$= \frac{1}{t^{\frac{n}{2}}} e^{\frac{i\pi}{2}} \frac{\eta(\Lambda)}{\sqrt{|\det\Lambda|}}$$

$\Sigma\text{sign}\lambda_i = \eta(\Lambda) = \text{signature}(\Lambda)$.

Now let us consider more general cases. Let $f : X \to \mathbf{R}$ be a smooth function with finitely many non-degenerate critical points P_α. At each P_α we choose coordinates x_i^α, such that P_α has coordinates $x_i^\alpha = 0$, and

$$f(x_i^\alpha) = f(P_\alpha) + \frac{1}{2!}\Sigma\lambda_{ij}^\alpha x_i x_j + \frac{1}{3!}\Sigma\lambda_{ijk}^\alpha x_i x_j x_k + ...$$

near P_α.

And we also wish to pick up coordinates x_i so that we have the measure $d\mu = \frac{dx_1...dx_n}{(2\pi)^{\frac{n}{2}}}$.

Then for large t we have $I(t) = \Sigma_\alpha I^\alpha(t)$ as asymptotic expansion series in $\frac{1}{t}$. Here

$$I^\alpha(t) = \int_{B^\alpha} \frac{dx_1...dx_n}{(2\pi)^{\frac{n}{2}}} \exp it(f(P_\alpha) + \frac{1}{2!}\Sigma\lambda_{ij}^\alpha x_i x_j + ...)$$

with B^α a small neighborhood of P_α.

To evaluate $I^\alpha(t)$, let $y_i = \sqrt{t}x_i$,

$$I^\alpha(t) = t^{-\frac{n}{2}} \int \frac{dy_1...dy_n}{(2\pi)^{\frac{n}{2}}} e^{itf(P_\alpha)} e^{\frac{1}{2}i\Sigma\lambda_{ij}^\alpha y_i y_j} e^{i(\frac{1}{3!\sqrt{t}}\lambda_{ijk}^\alpha y_i y_j y_k + \frac{1}{4!t}\lambda_{ijkl}^\alpha y_i y_j y_k y_l + ...)}$$

$$= t^{-\frac{n}{2}} e^{itf(P_\alpha)} \int \frac{dy_1...dy_n}{(2\pi)^{\frac{n}{2}}} e^{\frac{1}{2}i\Sigma\lambda^\alpha_{ij}y_iy_j}(1 + \frac{i}{3!\sqrt{t}}\lambda^\alpha_{ijk}y_iy_jy_k$$

$$+\frac{i}{4!t}\lambda^\alpha_{ijkl}y_iy_jy_ky_l + \frac{1}{2!}(\frac{i}{3!\sqrt{t}})^2(\lambda^\alpha_{ijk}y_iy_jy_k)^2 + O(\frac{1}{t^2})).$$

The integrals we need are (we will omit α subscript):

$$\int \frac{dy_1...dy_n}{(2\pi)^{\frac{n}{2}}} e^{\frac{1}{2}i\Sigma\lambda_{ij}y_iy_j} y_{s_1}...y_{s_r}$$

$$= (-i)^r \frac{\partial}{\partial a_{s_1}}...\frac{\partial}{\partial a_{s_r}} \int \frac{dy_1...dy_n}{(2\pi)^{\frac{n}{2}}} e^{\frac{1}{2}i\Sigma\lambda_{ij}y_iy_j} e^{i\Sigma a_s y_s}$$

And

$$\int \frac{dy_1...dy_n}{(2\pi)^{\frac{n}{2}}} e^{\frac{1}{2}i\Sigma\lambda_{ij}y_iy_j} e^{i\Sigma a_s y_s}$$

$$= \int \frac{dy_1...dy_n}{(2\pi)^{\frac{n}{2}}} e^{\frac{1}{2}i\Sigma\lambda_{ij}(y_i+(\lambda^{-1})_{ik}a_k)(y_j+(\lambda^{-1})_{jl}a_l)} e^{(-\frac{i}{2}\Sigma(\lambda^{-1})_{ij}a_ia_j}$$

$$= e^{\frac{i\pi}{2}} \frac{\eta(\Lambda_\alpha)}{\sqrt{\det \Lambda_\alpha}} e^{-\frac{i}{2}\Sigma(\lambda^{-1})_{ij}a_ia_j}.$$

Hence

$$\int \frac{dy_1..dy_n}{(2\pi)^{\frac{n}{2}}} e^{\frac{1}{2}i\Sigma\lambda_{ij}y_iy_j} y_{s_1}...y_{s_r}$$

$$= e^{\frac{i\pi}{2}} \frac{\eta(\Lambda_\alpha)}{\sqrt{\det \Lambda_\alpha}} (-i)^r (\frac{\partial}{\partial a_{s_1}}...\frac{\partial}{\partial a_{s_r}})|_{a=0} \exp(-\frac{i}{2}\Sigma(\lambda^{-1})_{ij}a_ia_j)).$$

Note that

$$\frac{\partial}{\partial a_s}\frac{\partial}{\partial a_t} \exp(-i(\lambda^{-1})_{ij}a_ia_j))|_{a=0} = -i(\lambda^{-1})_{st}.$$

So

$$\int \frac{dy_1 ... dy_n}{(2\pi)^{\frac{n}{2}}} e^{\frac{1}{2} i \Sigma \lambda_{ij} y_i y_j} y_{s_1} y_{s_2} y_{s_3} y_{s_4}$$

$$= (-i)^2 ((\lambda^{-1})_{s_1 s_2} (\lambda^{-1})_{s_3 s_4} + 2 \text{ more}).$$

And,

$$\lambda^{\alpha}_{ijk} \lambda^{\alpha}_{i'j'k'} \int \frac{dy_1 ... dy_n}{(2\pi)^{\frac{n}{2}}} e^{\frac{1}{2} i \Sigma \lambda_{ij} y_i y_j} y_i y_j y_k y_{i'} y_{j'} y_{k'}$$

$$= (-i)^3 e^{\frac{i\pi}{2}} \frac{\eta(\Lambda^\alpha)}{\sqrt{\det \Lambda^\alpha}} \lambda^{\alpha}_{ijk} \lambda^{\alpha}_{i'j'k'} ((\lambda^{-1})_{ii'} (\lambda^{-1})_{jj'} (\lambda^{-1})_{kk'}$$

$$+ (\lambda^{-1})_{ij} (\lambda^{-1})_{i'j'} (\lambda^{-1})_{kk'} + ...).$$

After Feynman this can be conveniently described by Feynman diagrams.

In the graph for each line put a factor of $(\lambda^{-1})_{ij}$. At each vertex put a factor of λ^{α}_{ijk} and so on. The numerical factor is $\frac{1}{N}$ with N the number of symmetries of that graph.

In summary we have got the asymptotic expansion series

$$I(t) = \int_X d\mu e^{itf}$$

$$= \Sigma_\alpha \frac{e^{\frac{i\pi}{2}}}{t^{\frac{\dim X}{2}}} e^{itf(P_\alpha)} \frac{\eta(\Lambda_\alpha)}{\sqrt{|\det \Lambda_\alpha|}} (1 + \Sigma_{n=1}^{\infty} \frac{C_{\alpha,n}}{t^n}),$$

where $C_{\alpha,n}$ are evaluated by Feynman diagrams.

5.2.2 *Integration on a sub-variety*

Now we consider the case that the set of critical points be a sub-manifold, or a sub-variety in general. In this case G acts on the set of critical points X. Let us assume that G acts on X freely. Then we have $\dim X = \dim X/G + \dim G$. Or in general let X_i be a component of X and G_i be the stabilizer of a generic $x \in X_i$, then $\dim X_i = \dim X_i/G + \dim G - \dim G_i$. And

$$\dim X - \dim X_i = \dim(X/G) - (\dim(X_i/G) - \dim G_i).$$

Here $\dim(X_i/G) - \dim G_i = \dim^*(X_i/G)$ is the formal dimension of X_i/G, it can be positive or negative. The contribution of X_i is

$$t^{-\frac{1}{2}\dim(X/G)} t^{-\frac{1}{2}\dim^*(X_i/G)}.$$

We have the formulae for $I(t)$ as follows.

$$I(t) = \Sigma_{X_i} t^{-\frac{1}{2}(\dim X - \dim X_i)} \int d\mu_i e^{\frac{i\pi}{2}} \frac{\eta(\Lambda_i)}{\sqrt{|\det \Lambda_i|}} (1 + \Sigma_n \frac{b_{n,i}}{t^n})$$

$d\mu_i$ is the transverse measure of X_i/G of the orbit space of a component X_i. Let μ_X be a measure on X, we want to construct a measure $\mu_{X/G}$ on X/G. For this we need to pick up a section $s : X/G \to X$. Let $W = s(X/G)$. The tangent bundle to W in X splits,

$$T_X W = T(W) + V(W).$$

On a vector space R a measure is a vector $x \in \Lambda^{\mathrm{Top}}(R)$. $\Lambda^{\mathrm{Top}}(T_X(W)) = \Lambda^{\mathrm{Top}}(T(W)) \times \Lambda^{\mathrm{Top}}(V(W))$.

The G action gives a map $\phi : \mathcal{G} \to V$, $\mathcal{G} = \mathrm{Lie}(G)$. Fix a measure on $\mathcal{G}, x \in \Lambda^{\mathrm{Top}}(\mathcal{G})$, then $\phi(x) \in \Lambda^{\mathrm{Top}}(V) = \mu_V$. The desired measure μ_W is defined by $\mu_X = \mu_V \times \mu_W$.

Let s be defined locally by equations $h_1 = ... = h_m = 0$. Then

$$1 = \int_G dg \Pi_{i=1}^n \delta(h_i(gx)) |\det(\partial_a h_i)|.$$

Here ∂a is the left invariant vector field on $\mathcal{G}$, $x \in$ fiber over X/G. There is a unique g such that $gx \in W = s(X/G)$. And

$$I(t) = \int_X d\mu_X e^{itf} = \int_G dg \int_X d\mu_X e^{itf(x)} \delta(h_i(gx)) |\det(\partial_a h_i)|.$$

Recall $\frac{1}{t}\delta(x) = \int_{-\infty}^{\infty} \frac{d\phi}{2\pi} e^{i\phi x t}$. Let ϕ_i be linear functions on $\mathcal{G}^*$. And we have

$$I(t) = \int_X d\mu_X e^{itf}$$

$$= \mathrm{vol}(G) t^{\dim G} \int_{\mathcal{G}^*} d\phi_1 ... d\phi_m \int_X d\mu_X e^{itf(x)} e^{it\Sigma\phi_i h_i(x)} \delta(h_i(gx)) |\det(\partial_a h_i)|.$$

For the materials above please see Faddeev's book for more details [Faddeev].

5.3 Partition function and topological invariants

Our goal is to get topological invariants for a three manifold M^3. Let G be a compact Lie group, $k \in \mathbf{Z}^+$. Let $A \in \mathcal{A}$ be a connection and $\mathcal{A}$ be the space of connections. We knew that the gauge group $\mathcal{G}$ acts on the space of connections. We denote $\mathcal{A}/\mathcal{G}$ as the space of gauge equivalent connections. The Chern-Simons Lagrangian is

$$\mathcal{L} = \frac{1}{4\pi} \int_M \mathrm{Tr}(A \wedge dA + \frac{2}{3} A \wedge A \wedge A).$$

Because for two gauge equivalent connections $\mathcal{L}$ only differs by an integer, $\exp(ik\mathcal{L})$ is well defined for gauge equivalent connections. We define topological invariants through the partition function:

$$Z_{G,k}(M) = \int_{\mathcal{A}/\mathcal{G}} \mathcal{D}A \exp ik\mathcal{L}.$$

It can also be generalized as follows: given a link $L = \cup_i C_i$, where C_i are knots with $C_i \cap C_j = \phi, i \neq j$. For each i pick an irreducible representation R_i of G and we introduce "holonomy"

$$W_{R_i}(C_i) = \mathrm{Tr}_{R_i} P \exp \int_{C_i} A.$$

We then generalize the partition function to be

$$Z_{G,k}(M; C_i, R_i) = \int_{\mathcal{A}/\mathcal{G}} \mathcal{D}A \exp(ik\mathcal{L}) \prod_i W_{R_i}(C_i).$$

It does not depend on metrics on M^3. It only depends on orientation of M^3 and on orientation of the link L for the path integral is integrated over the space of connections.

The problem is how to calculate $Z_{G,k}(M)$. We first rescale the field A to normalize the quadratic term and the coupling constant is then changed to $\frac{2}{3}k^{-\frac{1}{2}}$. We now consider how to construct perturbation series of the above path integral in terms of $k^{-\frac{1}{2}}$. Another method is to evaluate the integral via surgery explained in the last chapter.

5.3.1 *Gauge fixing and Faddeev-Popov ghosts*

Since the Chern-Simons Lagrangian is gauge invariant the integral is really over gauge equivalent fields, i.e. over $\mathcal{A}/\mathcal{G}$. To imitate this in gauge theory we need a local section:

$$s : \mathcal{A}/\mathcal{G} \to \mathcal{A}.$$

To get a section we pick a metric on M. The metric induces a Hodge operator $*$: k forms $\to$ 3-k forms. Let $A = A_0 + B$, where A_0 is a critical point of $\mathcal{L}$, i.e. A_0 is a flat connection. We only consider a simple case where we have an isolated critical point, or it amounts to be the same, $H^0 = H^1 = 0$. Then the section is defined by requiring that B obey the differential equation: $D{*}B = 0$. Here $D = d + A_0$ is the covariant derivative. An infinitesimal gauge transformation α transforms B into $B - D\alpha$. The constraint is then satisfied by solving equation $D * D\alpha = D * B$, or, $*D*$

$D\alpha = *D * B$. This gives a local section s. We have BRST transformation laws:

$$\delta A_i^a = -D_i c^a = -(\partial_i c^a + f_{bc}^a A_i^b c^c),$$

$$\delta c^a = \frac{1}{2} f_{bc}^a c^b c^c.$$

Here c is a zero form. The first identity above expresses the variation of a connection under an infinitesinal gauge transformation. The second equation states that local gauge transformation satisfies Maurer-Cartan equations.

The partition function is now:

$$Z = Z_{G,k;A_0}(M)$$

$$= \int \mathcal{D}B\mathcal{D}\phi\mathcal{D}c e^{\frac{ik}{2\pi}L_{CS}(A_0)} e^{\frac{ik}{2\pi}\int_M \mathrm{Tr}(B\wedge DB)+\mathrm{Tr}\phi(*D*B)+\frac{2}{3}\mathrm{Tr}(B\wedge B\wedge B+\bar{c}D_i D^i c)}.$$

ϕ are Lie algebra valued 3-forms. It is the Lagrangian multiplier to the constraint $D * B = 0$.

To pass from the integral over $\mathcal{A}/\mathcal{G}$ to the integral over $\mathcal{A}$, there is a change of measure. The field c is to rescale the measure of the gauge group by the determinant of the map $\alpha \to *D * D\alpha, \det(*D * D) = \det D^* D$, where $D^* = *D*$. It is included in the partition function as follows:

$$\int \mathcal{D}c \exp(i\mathrm{Tr}\bar{c}D_i D^i c) = \det(D^* D).$$

The fields $\bar{c}, \phi$ satisfy $\delta\bar{c} = 0, \delta\phi = 0$. Here $\bar{c}$ is a three form, ϕ is a zero form, $\Delta = D^* D$ is the Laplacian operator.

The BRST operator δ satisfies $\delta^2 = 0$. So δ induces a cohomology. The zero order cohomology gives gauge equivalent classes. That is why we say that BRST operator selects gauge invariant objects.

In this setting, the new Lagrangian can be written in another form, $\mathcal{L}' = \mathcal{L} - \delta V$. Choose V so that the kinectic energy of $\mathcal{L}'$ is non-degenerate. For example, we can choose V as follows.

$$V = \frac{k}{2\pi}\int_M d\mu \mathrm{Tr}\bar{c} * D * B.$$

Then we have

$$\delta V = \frac{k}{2\pi}\int_M d\mu \mathrm{Tr}\delta\bar{c} * D * B + Tr\bar{c} * D * \delta B$$

$$= \frac{k}{2\pi}\int_M d\mu \mathrm{Tr}(i\phi * D * B - \bar{c} * D * Dc).$$

5.3.2 *The leading term*

The leading term is given by:

$$\int \mathcal{D}B\mathcal{D}\phi \exp(i\int_M \mathrm{Tr}(B \wedge DB + \phi * D * B - \bar{c} * D * Dc)).$$

The functional here is quadratic in $H = (B, \phi)$. We have

$$\mathrm{Tr}(B \wedge DB + \phi * D * B)) = (H, L_- H),$$

where $L = (*D + D*)$ is a twisted Dirac operator acting on forms. It maps even forms to even forms and odd forms to odd forms and L_- is the restriction of L to odd forms, $(,)$ is the natural inner product

$$(\phi, \phi^{'}) = \int \mathrm{Tr}(\phi \wedge *\phi^{'}).$$

It can be seen as follows. $*DB, D * \phi$ are one forms, $D * B$ is a 3 form. So we have

$$(H, L_- H) = \int_M \mathrm{Tr}(B \wedge DB) + \int_M \mathrm{Tr}(B \wedge *D * \phi) + \int \mathrm{Tr}(\phi * D * B).$$

We know that $(\phi, \phi^{'}) = (\phi^{'}, \phi)$. When M is a closed manifold, by applying Stokes theorem, we have

$$\int_M \mathrm{Tr}(B \wedge *D * \phi) = \int_M \mathrm{Tr}(\phi * D * B).$$

Let us calculate the path integral:

$$\int \mathcal{D}H\mathcal{D}\bar{c}\mathcal{D}c \exp(-\frac{i}{2}(H, L_- H) + (\bar{c}, \Delta c)).$$

We first choose orthonormal eigenfunctions of those two operators. We have:

$$\Delta\psi_i = u_i\psi_i, (\psi_i, \psi_j) = \delta_{ij};$$

$$L_-\xi_i = v_i\xi_i, (\xi_i, \xi_j) = \delta_{ij}.$$

Let $c = \Sigma_i c_i\psi_i, \bar{c} = \Sigma_i \bar{c}_i\bar{\psi}_i, H = \Sigma_i h_i\xi_i$.

$$\mathcal{D}\bar{c}\mathcal{D}c = \Pi_i d\bar{c}_i dc_i, \mathcal{D}H = \Pi_i \frac{dh_i}{\sqrt{2\pi}}.$$

We then have

$$Z = \Sigma_{A^0} \frac{1}{\sharp Z(G)} e^{ikI(A^0)} \int \Pi_i \frac{dh_i}{\sqrt{2\pi}} \Pi_j d\bar{c}_j dc_j \exp(-\frac{i}{2}\Sigma_i v_i h_i^2 + \Sigma_j u_j \bar{c}_j c_j).$$

Notice that:

$$\int d\bar{c}dc \exp(u\bar{c}c) = \int d\bar{c}dc(1 + u\bar{c}c) = u,$$

$$\int_{-\infty}^{\infty} \frac{dx}{\sqrt{2\pi}} e^{-ivx^2/2} = \frac{e^{i\pi \mathrm{sign}(v)/4}}{\sqrt{|v|}}.$$

So we have

$$Z = \frac{1}{\sharp Z(G)} e^{ikI(A^0)} \frac{\det \Delta e^{-i\pi\eta(L_-)/4}}{\sqrt{|\det L_-|}}.$$

We have $\det\Delta = \Pi_i u_i$, $\eta(L_-) = \Sigma_j \text{sign} v_j$. The determinant of Laplacian operator Δ can be regularized by using zeta function, $\det\Delta = \exp(-\zeta'(0))$, $\zeta(s) = \Sigma_i u_i^{-s}$. The η invariant can be regularized in a similar way, $\eta(L_-) = \eta(L_-, 0)$, $\eta(L_-, s) = \Sigma_j \text{sign} v_j |v_j|^{-s}$.

It is shown by Schwarz that $\frac{\det(\Delta)}{\sqrt{\det(L_-)}}$ is a topological invariant. Its absolute value is the same as Ray-Singer analytic torsion, and the imaginary part gives the η invariant,

$$\frac{1}{\sqrt{\det L_-}} = \frac{1}{\sqrt{|\det L_-|}} \exp(\frac{i\pi}{2}\eta(A^\alpha)).$$

Here $\eta(A^\alpha) = \frac{1}{2}\lim_{s\to 0}\Sigma_i \text{sign} v_i |v_i|^{-s}$, or formally $\eta(A^\alpha) = \frac{1}{2}\Sigma_i \text{sign} v_i$.

The η invariant is related to Chern-Simons functional by $\frac{1}{2}(\eta(A^\alpha) - \eta(0)) = \frac{c_2(G)}{2\pi} I(A^\alpha)$, where $I(A^\alpha)$ is the Chern-Simons action of A^α and $\eta(0)$ is the η invariant of the trivial gauge field $A = 0$. $c_2(G)$ is the Casimir for G. For example, we have $c_2(SU(2)) = 2N$.

Finally, we have the formula for the leading term as well as the formula for Abelian Chern-Simons gauge theory,

$$Z = e^{\frac{i\pi\eta(0)}{2}} \Sigma_\alpha e^{\frac{i(k+N)}{2} I(A^\alpha)} T_\alpha,$$

$T_\alpha = \frac{\det(\Delta)}{|\sqrt{\det(L_-)}|}$ is the torsion invariant of A^α.

To define self-linking integrals and its non-Abelian generalizations we need a framing of the knot so that one can consider the case x, y, z on the knot to coincide. For a given knot K, consider a normal vector field on K. Imagine moving the knot slightly in the direction of the vector field we then get another knot K'. The self-linking number of K can be defined as the linking number of K and K'. Such a definition depends on a choice of a normal vector field which is called a framing of a knot. It is easy to see that the choice only depends on homotopy classes of the normal vector field.

The changing of framing is measured by an integer s. It likes to twist K' around K, s times. The partition function changes as $Z \to \exp(2\pi i s h)Z$, where h is the conformal weight. So we know how it changes under different framing. In this sense, a choice of framing does not trouble us.

For Abelian theory we have $h_i = \frac{n_i^2}{2k}$. The Dehn twist around a marked point acts as $e^{2\pi i h_i}$.

In general, there exists a constant $c = c(G, k)$, the central charge of Virasoro, such that if framing of M is changed by r units then the partition function transforms as

$$Z(M) \to e^{2\pi irc} Z(M).$$

5.3.3 *Wilson line and link invariants*

Let L be a link with several embedded circles C_i in M^3. For each circle we associate a representation R_i. The partition integral is then modified to

$$Z(M) = \int \mathcal{D}A e^{\frac{ik}{4\pi}} \int_M \mathrm{Tr}(A \wedge dA + \frac{2}{3} A \wedge A \wedge A) \prod_i \mathrm{Tr}_{R_i} P \exp \int_{C_i} A.$$

Let A_0 be a critical point and let $A = A_0 + B$ be perturbations around the critical point. Then we can expand

$$\mathrm{Tr}_{R_i} P \exp \int_{C_i} A = \mathrm{Tr}_{R_i} P \exp \int_{C_i} (A_0 + B)$$

$$= \mathrm{Tr}_{R_i} P \exp \int_{C_i} A_0 + \int_{C_i} \mathrm{Tr} B\Gamma + \int_{C_i \times C_i} \mathrm{Tr} B\Gamma_1 B\Gamma_2 + ...,$$

where Γ corresponds to Green functions of the quardratic form by L_-.

Now suppose that $G = U(1)$. In this case we have R_n is corresponding to $n, n \in \mathbf{Z}$. Let A be an one-form. Then we have

$$\exp in \int_{C_i} A = \exp in \int_{C_i} (A_0 + B)$$

$$= \exp(in \int_{C_i} A_0 (1 + in \int_{C_i} B + \frac{1}{2}((in)^2 \int_{C_i} B)^2 + ...)).$$

For a term of order two, it corresponds to a wave-line. It then corresponds to $\frac{1}{L_-} = G(x, y)$, a $(1, 1)$ form on $M \times M$. For the flat metric on $\mathbf{R}^3, G_{ij}(x, y) = \frac{\epsilon_{ijk}(x-y)_k}{|x-y|^3}$.

$$\frac{1}{Z_0}\int \mathcal{D}B\mathcal{D}\phi e^{\frac{ik}{4\pi}\int (B,\phi)L_-(B,\phi)^{tr}} \int_{C_1} B \int_{C_2} B$$

$$= -\frac{4\pi i}{k}\int_{C_1\times C_2} G(x,y)$$

$$= -\frac{4\pi i}{k} \text{ Link } (C_1, C_2).$$

In general one can expand holonomy of Wilson line in terms of iterated path integrals from K. T. Chen [Chen]:

$$P\exp \int_C A = 1 + \Sigma_{m=0}^{\infty} \int_{0\le t_1 \le \ldots \le t_m \le 1} (C^* A(t_m))\ldots(C^* A(t_1)).$$

This was used by D. Bar-Natan to establish a relation of Feynman diagrams in Chern-Simons theory with that of Vassiliev invariants [Bar-Natan].

5.4 A brief introduction on renormalization of Chern-Simons theory

Since the seminal work of Witten [Witten] on Chern-Simons theory such a theory becomes a model of topological quantum field theory. Witten's approach is non-perturbative by making a connection with conformal field theory. In this section we continue the work to construct the theory through renormalized perturbation series. Classically the Chern-Simons action enjoys two kinds of symmetries, namely gauge symmetry and diffeomorphism symmetry. It is remarkable that those two symmetries survive in the quantum level for some regulization schemes. This has been explored by several authors [Blasi and Collina], [Chen-Semenoff-Wu]. They found that there are nice cancellations among those divergent terms. The cancellations resemble those in $N = 2$ supersymmetry theories. The remaining convergent terms are the same as if the coupling constant for the Chern-Simons terms has been shifted by $c_2(G)$.

5.4.1 *A regulization scheme*

To quantize Chern-Simons theory is to evaluate the path integral:

$$Z = \int \mathcal{D}A e^{\frac{ik}{2\pi} I_{CS}}.$$

The integral is over the space of gauge equivalent fields. The construction of perturbation series depends on a choice of regulization scheme. If we use high derivative regulization scheme in the Lorentz gauge, we add following terms to the original Lagrangian,

$$I_{tot} = I_{CS} + I_{YM} + I_{gf} + I_{gh}$$

where we have

$$I_{YM} = -\frac{1}{2e^2} \int d^3x F_{\mu\nu} F^{\mu\nu},$$

$$I_{gf} = -\frac{1}{\alpha e^2} \int d^3x (\partial^\mu A_\mu)^2,$$

$$I_{gh} = \int d^3x (\partial^\mu \bar{c} D_\mu c).$$

The term I_{gf} is the Lagrangian multiplier for the constraints $d^* A = 0$. The term I_{gh} is the determinant of the Laplace operator. The original path integral is then replaced by:

$$Z' = \int \mathcal{D}A \mathcal{D}c \mathcal{D}\bar{c} e^{I_{tot}}.$$

We can verify that I_{tot} still respect BRST symmetry and diffeomorphism symmetry. The BRST symmetry is:

$$\delta A^a_\mu = -(\partial_\mu c^a + f^{abc} A^c_\mu c^b),$$

$$\delta c^a = \frac{1}{2} c^b c^c, \delta \bar{c} = \phi, \delta \phi = 0.$$

Assuming we have an infinitesimal diffeomorphism $x^\sigma \to x^\sigma - \lambda^\sigma$, it acts on all the fields above and the diffeomorphism symmetry is:

$$\delta A^a_\mu = \lambda^\sigma \partial_\sigma A^a_\mu + \partial_\mu \lambda^\sigma A^a_\sigma,$$

$$\delta c^a = \lambda^\sigma \partial_\sigma c^a,$$

$$\delta \zeta^{\mu\nu} = \lambda^\sigma \partial_\sigma \zeta^{\mu\nu} - \partial_\sigma \lambda^\mu \zeta^{\sigma\nu} - \partial_\sigma \lambda^\nu \zeta^{\mu\sigma} + \frac{2}{3} \partial_\sigma \lambda^\sigma \zeta^{\mu\nu},$$

$$\delta \phi^a = \lambda^\sigma \partial_\sigma \phi^a + \frac{1}{3} \partial_\sigma \lambda^\sigma \phi^a,$$

$$\delta \bar{c}^a = \lambda^\sigma \partial_\sigma \bar{c}^a + \frac{1}{3} \partial_\sigma \lambda^\sigma \bar{c}^a.$$

To verify that I_{tot} respects diffeomorphism symmetry it is crucial to observe that d commutes with diffeomorphisms.

5.4.2 *The Feynman rules*

The Feynman rules are:

1. The gluon propagator:

$$G_1(x, y) = \frac{1}{L_-}.$$

2. The ghost propagator:

$$G_2(x, y) = \frac{1}{D^* D + D D^*}.$$

3. The ghost-gluon vertex coming from $\bar{c} c A$.
4. The three gluon vertex:

Fig. 5.1 Feynman diagrams

$$A_i^j \wedge A_j^k \wedge A_k^i.$$

5. The four gluon vertex coming from the Yang-Mills term.

Given the Feymann rules above, one can construct Feymann diagrams. We can organize the diagrams according to the number of loops. Here are some of the one-loop divegent Feynman diagrams and they cancel each other. The solid line corresponds to a gluon propagator and the dashed line corresponds to a ghost propagator. We have three kinds of vertext here, the three gluon vertex, the gluon-ghost-ghost vertex and the four gluon vertex.

We found the following observations crucial.

1. Divergent diagrams cancelled from each other because of gluon-ghost symmetry. For any divergent diagram we can always find a loop which contains both gluon and ghost propagator. If we exchange gluon propagators with ghost propagators we obtain another diagram. The resulting diagram is also admissible with a minus sign and it canceled the original one. This is true because of the special structure of the Feymann rules.

2. The remaining convergent diagrams is the same as if the coupling constant of the Chern-Simons theory shifted by $c_2(G)$. By taking a limit we restore the two symmetries at the quantum level. The only term which respect the two symmetries is the Chern-Simons action itself. And the only

counter term which survive after taking the limit is the Chern-Simons term. This expalins why we can only have the coupling constant shifted.

Chapter 6

Topological Sigma Model and Localization

6.1 Constructing knot invariants from open string theory

6.1.1 *Introduction*

A knot is an embedding of a circle into a three-dimensional space, say $K : S^1 \to S^3$. Two knots K_1, K_2 are equivalent, if their complements in S^3 are homeomorphic. So when we talk about a knot, we are really talking about a three manifold, i.e. the complement of a circle in S^3.

There are two natural ways to construct knot invariants. One way is what we have been doing based on Chern-Simons-Witten quantum field theory. For any three manifold consider a principle bundle with gauge group G and all gauge fields indexed by the three manifold. We then average over the space of all gauge fields with Chern-Simons Lagrangian as their weight. If such an average make sense it would give topological invariants. In previous chapters we explained how to do this from several points of views.

There is another natural construction. For a given knot $K : S^1 \to S^3$, we consider all embeddings $\Phi : K \to S^3$ in the same topological class. We then do an average with respect to all such embeddings. Again if the average make sense then it would give topological invariants because they do not depend on particular embeddings. Witten proposed to do this via open string theory, or topological sigma models. We shall give a brief introduction here. For more details, please see Witten's paper [Witten-string]. For the latest development please see Vafa's recent paper [Vafa].

6.1.2 *A topological sigma model*

Witten's idea is to complexify the embedding. He considers a knot as a boundary component of a surface Σ and the target space be T^*M^3. Then he considers all mappings $\Phi : \Sigma \to T^*M^3$. There is a natural symplectic structure on the target space $\omega = d\alpha, \alpha = \Sigma q_i dp_i$. We have a Lagrangian submanifold $M^3 \subset T^*M^3$, i.e. $\omega|_M = 0$. Choose an almost complex structure J on T^*M. Let $g = \omega(.,J.) > 0$ be a metric on T^*M and g is of (1,1) type, $g_{ij} = 0, g_{\bar{i}\bar{j}} = 0, g_{i\bar{j}} = g_{\bar{j}i} = -i\omega_{i\bar{j}}$. We then consider all maps $\Phi : \Sigma \to T^*M$ with $\Phi(\partial\Sigma) \subset M$.

To construct a topological model, we also need to include fermions. Let $\Phi : \Sigma \to X$ be a given map. We then have two kinds of fermions $\chi \in \Gamma(\Sigma, \Phi^*(TX)), \psi \in \Omega^1(\Sigma, \Phi^*(TX))$. Since $TX = T^{(1,0)}X \oplus T^{(0,1)}X$, we have $\psi = \psi^i + \psi^{\bar{i}}$.

There are symmetries of those fields due to variations of the mapping and the derived variations on fermions and we shall have a Lagrangian to be invariant under those variations. We have

$$\delta\Phi^I = i\alpha\chi^I, \delta\chi^I = 0,$$

$$\delta\psi^{\bar{i}} = -\alpha\partial_z\Phi^{\bar{i}} - i\alpha\chi^{\bar{j}}\Gamma^{\bar{i}}_{\bar{j}\bar{m}}\psi^{\bar{m}}_z,$$

$$\delta\psi^{\bar{i}}_z = \alpha\partial_{\bar{z}}\Phi^i - i\alpha\chi^j\Gamma^i_{jm}\psi^m_{\bar{z}}.$$

Those are BRST transformations. If we denote Λ for the collection of fields, we have $\delta\Lambda = -i\alpha\{Q, \Lambda\}$, and $Q^2 = 0$.

We can choose Lagrangian to be $L = i\{Q, V\} + t\int_\Sigma \Phi^*\omega$. For a good theory, we need to choose V such that the kinetic part of L is nondegenerate. For example, one can choose $V = t\int_\Sigma d^2z g_{i\bar{j}}(\psi^{\bar{i}}_z\partial_{\bar{z}}\Phi^j + \partial_z\Phi^{\bar{i}}\psi^j_{\bar{z}})$.

We then have a supersymmetric Lagrangian to be

$$L = 2t\int_\Sigma d^2z(g_{IJ}\partial_z\Phi^I\partial_{\bar{z}}\Phi^J + i\psi D_z\chi + i\psi D_{\bar{z}}\chi - R\psi\bar{\psi}\chi\bar{\chi}) + t\int_\Sigma \Phi^*(\omega),$$

$\int_\Sigma \Phi^*(\omega) = 2\pi n$ is the degree of the mapping Φ. It is clearly invariant under Q.

To quantize the model is to calculate the path integral:

$$\int \mathcal{D}\Phi e^{iL}.$$

Witten argued that this integral makes sense as an open string theory. It is a topological field theory. The theory does not depend on t, so the perturbation series in t is exact. Usually topological sigma models can be understood by using localization principle. In this case the moduli space is degenerate. However it can be understood by using triangulation of Teichmuller space. Witten also argued that it is equivalent to large N expansion of the Chern-Simons gauge theory.

6.1.3 *Localization principle*

Let us give a short introduction of localization principle. We will give more detail explanation in the following sections.

Let us first consider the finite dimensional case. Let G be a Lie group and g its Lie algebra, G acts on a symplectic space preserving the symplectic structure. Let $\Omega^*(X)$ be the space of de Rham complex. Let us consider the space $\Omega^*(X) \otimes \mathrm{Fun}(g)$. Let $D = d - i_V$, where V is the Killing vector field. A form $\alpha \in \Omega^*(X) \otimes \mathrm{Fun}(g)$ such that $D^2\alpha = 0$ is called an equivariant differential form. The induced D cohomology is called equivariant cohomology.

Let α be any equivariant differential form, then we have:

$$\int_X \alpha = \int_X \alpha e^{tD\lambda}$$

$$= \Sigma_\sigma \text{ local expressions}$$

where $X_\sigma = \{\lambda(V_i) = 0)\}$ are components of the fixed point set of the G action, and λ is any one form. It is true because we can take a large t limit and the integral localize to the fixed point set of the G action. We will justify this in the following sections for several interesting cases.

However in the case of field theory, we will be in the infinite dimensional setting. In the case of open string theory the space X will be the space of fields. The group G will be replaced by the group of symmetries acting on the fields. The operator D will be replaced by the BRST operator Q. We then need to evaluate the integral

$$\int \mathcal{D}\Phi\mathcal{D}\chi\mathcal{D}\psi e^{it\int\{Q,V\}}\Pi_a\mathcal{O}_{H_a}(P_a).$$

The equivariant differential form is now replaced by $\mathcal{O}_{H_a}(P_a)$, where H_a are submanifolds in M and $\mathcal{O}_{H_a}(P_a)$ is the Poincaré dual in $H^*(M,\mathbf{R})$.

Again the path integral is localized to the fixed point set. The fixed point set is now a vector bundle (given by fermions) over the space of pseudo holomorphic maps (given by bosons). Pseudo holomorphic maps are indexed by the degree of the maps. For large t, the operator Q can be identified with the usual differential operator d. The Q-cohomology is then the same as the usual cohomology $H^*(M,\mathrm{End}(E))$. The correlation functions are then given by:

$$< \Pi_{a=1}^{s}\mathcal{O}_{H_a}(P_a) >= \Sigma_{n=0}^{\infty}e^{-2\pi nt}\int_{\mathcal{M}_n}\chi(\mathcal{V}),$$

where $\chi(\mathcal{V})$ is the Euler class of the bundle $\mathcal{V}$ over $\mathcal{M}$.

The above is usually called A model. There is another variant of topological sigma models called B-model.

6.1.4 *Large N expansion of Chern-Simons gauge theory*

In the perturbative expansion for Chern-Simons gauge theory, the main interaction term is:

$$\mathrm{Tr}(A\wedge A\wedge A) = A_j^i\wedge A_k^j\wedge A_i^k.$$

We then have many Feynman diagrams. t'Hooft used open Riemann surface to express those Feynman diagrams. For each line one replaces it by a strip, then one has a surface with many holes. Assuming we have h boundary components and r loops, then the genus of the surface is $g = \frac{r-h+1}{2}$. The free energy of the $\frac{1}{N}$ expansion of Chern-Simons theory is

$$F = \log Z = \Sigma_{g=0,h=1}C_{g,h}N^hk^{2g-2+h}.$$

where $k = \frac{2\pi}{k+N}$.

Let $Z_{g,h}$ be the partition function of an open string theory on a world-sheet with g handles and h boundaries. Witten argued that

$$C_{g,h} = Z_{g,h}.$$

The claim can be justified by considering triangulations of Teichmuller space based on fatgraph method. For references, see [Witten-string]. Most recently Vafa et al. proposed using closed string theory to construct knot invariants, see [Vafa].

6.2 Equivariant cohomology and localization

6.2.1 *Equivariant cohomology*

Let X be a manifold, G a Lie group, G acts on X, i.e. we have

$$\rho : G \to \mathrm{Diff}(X).$$

By taking derivatives, we have

$$d\rho : \mathbf{g} \to \mathrm{Vect}(X).$$

ρ induces an action on the de Rham complex $\Omega^*(X)$, by $(g, \omega) \to g^*(\omega)$. We shall enlarge the space $\Omega^*(X)$ into $\Omega^*(X) \otimes \mathrm{Fun}(\mathbf{g})$. Here $\mathrm{Fun}(\mathbf{g})$ consists of polynomial functions on Lie algebra $\mathbf{g}$. $\omega \in \Omega^*(X) \otimes \mathrm{Fun}(\mathbf{g})$ is a differential form on X with coefficients in $\mathrm{Fun}(\mathbf{g})$. Again ρ induces an action on $\Omega^*(X) \otimes \mathrm{Fun}(\mathbf{g})$. By taking derivative, we see that infinitesimally, the action is given by $\omega \to L_V\omega, V \in d\rho(\mathbf{g})$. Here $L_V = di_V + i_V d$ is the Lie derivative. Those forms ω satisfying $L_V\omega = 0, V \in d\rho(\mathbf{g})$ are called equivariant differential forms. They are also elements in G-invariant subspace of $\Omega^*(X) \otimes \mathrm{Fun}(\mathbf{g}), \Omega^*_G(X) = (\Omega^*(X) \otimes \mathrm{Fun}(\mathbf{g}))^G$.

We define a twisted differential operator $D = d + i_V$ on $\Omega^*_G(X) = (\Omega^*(X) \otimes \mathrm{Fun}(\mathbf{g}))^G$. We have $D^2 = di_V + i_V d = L_V$. So $D^2 = 0$ on $\Omega^*_G(X) = (\Omega^*(X) \otimes \mathrm{Fun}(\mathbf{g}))^G$. Thus D induces a cohomology on the complex $\Omega^*_G(X), H^*_G(X) = \mathrm{Ker}D/\mathrm{Image}D$ is called equivariant cohomology.

Example (Atiyah and Bott): Let T^n be a torus which acts on X freely, then $H_T^*(X) = H^*(X/T)$.

In general we may have fixed points for the action. Let Ω be a top equivariant differential form. We will find that the integration of Ω usually reduce to an integral over the set of fixed points.

6.2.2 *Localization, finite dimensional case*

Witten [Witten-localization] defined equivariant integration as

$$\int \alpha = \frac{1}{\mathrm{vol}(G)} \int_{\mathbf{g}\times X} \frac{d\phi_1 ... d\phi_s}{(2\pi)^s} \alpha \exp(-\frac{\epsilon}{2}(\phi,\phi)).$$

It is easy to check that $\int D\beta = 0, \beta \in \Omega_G^*(X)$. So we have a map:

$$\int : H_G^*(X) \to \mathbf{C}.$$

If we only integrate the X part, we would have

$$\int : H_G^*(X) \to H_G^*(pt).$$

For any one form $\lambda \in \Omega_G^*(X)$, we have

$$\int_X \alpha = \int_X \alpha e^{tD\lambda}.$$

For $e^{tD\lambda} = 1 + tD\lambda, \int \alpha D\lambda = 0$. Picking an orthonormal basis T_a of g, let $V(\phi) = \Sigma_a \phi^a V_a, V_a$ be the vector field representing T_a. Then we have

$$\int_X \alpha \frac{1}{\mathrm{vol}(G)} \int_{\mathbf{g}\times X} \frac{d\phi_1 ... d\phi_s}{(2\pi)^s} \alpha \exp(td\lambda - it\sigma_a \phi^a \lambda(V_a) - \frac{\epsilon}{2}\sigma(\phi^a)^2)$$

$$= \frac{1}{\mathrm{vol}(G)(2\pi\epsilon)^{s/2}} \int_X \alpha \exp(td\lambda - \frac{t^2}{2\epsilon}\Sigma_a(\lambda(V_a))^2).$$

As $t \to \infty$, outside a neighborhood of $X' = \{x|\lambda(V_a) = 0, a = 1, ..., s\}$, the integral has contribution e^{-at^2}. Let $X' = \cup_{\sigma \in S} X_\sigma, X_\sigma$ be its connected component. Then we have

$$\int_X \alpha = \Sigma_{\sigma \in S} Z_\sigma.$$

Z_σ is determined by local data of α and the action near X_σ. We call the above formula the **localization principle**.

6.3 Atiyah-Bott's residue formula and Duistermaat-Heckman formula

6.3.1 *Complex case, Atiyah-Bott's residue formula*

Example (Atiyah and Bott): Consider a $U(1)$ action on X. Let V be the vector generated by the action. Let g be a G-invariant Riemannian metric, $\lambda = -g(V, .)$ be a one form. Then

$$\int_X \alpha = \int_X \alpha \exp(-td\lambda - tg(V, V)).$$

Let $t \to 0$, the above integral will localize at zeros of $g(V, V)$, i.e. the zeros of V. At an isolated point P of V, the Hessian of $g(V, V)$ is non-degenerate, large t limit is a Gaussian integral. This way, we derive Atiyah-Bott's fixed point formula.

In the following we derive the original residue formula of Bott for symmetries preserving complex structures.

Definition (Holomorphic Vector Bundle): A vector bundle $E \to M$ with a complex vector space as their fibers over a complex manifold M is called a holomorphic vector bundle. The complex structure on M induces an almost complex structure $J : T_{\mathbf{C}}M \hookleftarrow, J^2 = -Id$. Let $T_{\mathbf{C}}^{(0,1)}M = \text{Ker}(J + i), T_{\mathbf{C}}^{(1,0)}M = \text{Ker}(J - i)$. Then we have $T_{\mathbf{C}}M = T_{\mathbf{C}}^{(0,1)}M \oplus T_{\mathbf{C}}^{(1,0)}M$. We called vectors in $T_{\mathbf{C}}^{(0,1)}M$ type $(0, 1)$ vectors and vectors in $T_{\mathbf{C}}^{(1,0)}M$ type $(1, 0)$ vectors. Similar decompositions can be made for $T_{\mathbf{C}}^{*}M$ and their tensor products. We then have a notion of type (p, q) tensors.

Let $<,>$ be a Hermitian structure on E, i.e. a metric on E whose restriction on each fiber is a quadratic form of type $(1, 1)$. Let $\{s_i\}$ be local frames. Let $N = \{(s_i, s_j)\}$ be a matrix of inner product. We can write $d = \bar{\partial} + \bar{\bar{\partial}}.\theta(s) = \bar{\partial} N N^{-1}$ is of type $(1,0)$ is the connection associated with the metric. $\Omega(s) = \bar{\bar{\partial}}\theta(s)$ is its curvature. We have $\bar{\bar{\partial}}\Omega = 0$.

Let E be a complex vector bundle on a complex manifold M. Let $c_i(M) \in H^{2i}(M, Z)$ be the i-th Chern class of E. If $\Phi(c) = \Phi(c_1, ..., c_n)$ is a polynomial in the indeterminates c_i, then we have $\Phi(c(M)) \in H^*(M, C)$. We wish to evaluate the integral

$$\int_M \Phi(c(M)).$$

We called these characteristic numbers. If $w_a = c_1^{a_1} c_2^{a_2} ... c_s^{a_s}, a_1 + 2a_2 + ... + sa_s = k, k$ is called the weight of w_a. We see that only those monimals with $k = n$ contribute to the integral.

If $A : V \to V$ is an endmorphism of a finite dimensional vector space, let $\det(1 + \lambda A) = \sum_i \lambda^i c_i(A)$. We call $c_i(A)$ Chern classes for A, let $\Phi(A) = \Phi\{c_1(A), ..., c_n(A)\}$.

Let X be a vector field preserving the complex structure on M. Let $L_X = i_X d + di_X$ be its Lie derivative. Restricted to zero's of X, it induces a map $L_p(X) = L_X | TM, p \in \text{zero}(X)$.

Here is Bott's residue formula which was soon generalized to a much more general case by Atiyah and Bott.

Theorem: Let X be a nondegenerate vector field that preserves a complex structure on M. Then for every polynomial $\Phi(c_1, ..., c_n)$ of weight not greater than n, we have

$$\Sigma_p \Phi(L)/c_n(L) = \Phi(M).$$

Here p ranges over zeros of X and $L = L_X | T_p(M), c_n(L) = \det L$.

In the proof, Bott expresses $\Phi(M)$ as an exact form away from zeros of X. So the integral is localized to zeros of X which then can be evaluated by local considerations. For a complete elegant proof see: R. Bott, **Michigan Journal of Math.** 14 (1967) 231-244.

6.3.2 *Symplectic case, Duistermaat-Heckman formula*

Let X be a $2n$ dimensional compact symplectic manifold with symplectic form ω. Suppose that the group $U(1)$ acts symplectically on X, the action being generated by a vector field V. The action is said to be Hamiltonian if there is a function H on X such that $dH = -j_V\omega$.

We wish to evaluate the following integral:

$$\int_X \frac{\omega^n}{n!}e^{-\beta H}.$$

The Duistermaat-Heckman formula asserts that this integral is given by the semi-classical approximation. If the critical points are isolated points P_i, then the formula is

$$\int_X \frac{\omega^n}{n!}e^{-\beta H} = \Sigma_i \frac{e^{-\beta H(P_i)}}{\beta^n e(P_i)},$$

where $e(P_i)$ is the product of the weights of the circle action in the tangent space at P_i.

Proof: The G action on X has a moment map μ. We pick up an almost complex structure J on X such that ω is of type (1,1) and positive, i.e. $g(v,v) = \omega(v, Jv) > 0, v \neq 0$. Set $I = (\mu,\mu)$ and $\lambda = \frac{1}{2}J(dI)$. Critical points of μ are zeros of λ. $\mu^{-1}(0)$ is the set where μ achieve absolute minimum. Assuming that $\mu^{-1}(0)$ is a smooth manifold on which G acts freely, then $M = \mu^{-1}(0)/G$ is a smooth manifold with a natural symplectic structure. Contribution of $\mu^{-1}(0)$ is

$$Z(\mu^{-1}(0)) = \int_X \alpha = \frac{1}{\text{vol}(G)}\int \frac{d\phi_1...d\phi_s}{(2\pi)^s}\int_X \alpha \exp(tD\lambda - \frac{\epsilon}{2}(\phi,\phi)).$$

Let Y be an equivariant neighborhood of $\mu^{-1}(0)$. We have an equivariant projection $\pi : Y \to \mu^{-1}(0)/G$. Let $-(\phi,\phi)/2 = \pi^*(\Theta), \alpha = \pi^*(\alpha')$. Then,

$$Z(\mu^{-1}(0)) = \frac{1}{\text{vol}(G)}\int \frac{d\phi_1...d\phi_s}{(2\pi)^s}\int_Y \alpha \exp(tD\lambda + \epsilon\Theta).$$

By integrating over the fibers of $\mu^{-1}(0) \to \mu^{-1}(0)/G$, we have

$$Z(\mu^{-1}(0)) = \int_{\mu^{-1}(0)/G} \alpha' \exp(\epsilon\Theta).$$

Evaluating Gaussian integral gives the localization formula. For higher critical points of I, we have

$$Z(X_\sigma) \sim \exp(-\frac{I(X_\sigma)}{2\epsilon}).$$

All of them are exponentially small.

6.4 2D Yang-Mills theory by localization principle

Localization principle also applies to infinite dimensional settings. This principle is extremely useful in topological quantum field theory and in string theory. In the following we introduce Witten's treatment of two-dimensional Yang-Mills theory by using localization principle.

6.4.1 *Cohomological Yang-Mills field theory*

Let Σ be a surface, $E \to \Sigma$ a principle bundle with fiber G. The gauge group $\mathcal{G}$, consisting of gauge transformations $g : \Sigma \to G$, acts on the space of connections $\mathcal{A}$ by $(g, A) \to gAg^{-1} + dgg^{-1}$. We will apply localization principle for finite dimensional group action to the current case of infinite dimensional group action.

The tangent space of $\mathcal{G}$ is the space of one forms with values in the Lie algebra of G. Infinitesmally we have:

$$(u, A) \to D_A u = du + [u, A].$$

Here $u \in T\mathcal{G}$ is an infinitesimal gauge transformation. We can also write

$$\delta A_i = i\epsilon\psi_i,$$

$$\delta\psi_i = \epsilon D_i\phi = -\epsilon(\partial_i\phi + [A_i, \phi]),$$

$$\delta\phi = 0.$$

Here $\psi_i \in \Gamma(\Sigma, K \otimes \mathbf{g})$ is an anti-commuting one form with values in the adjoint representation of G. It is often called a fermion. $\phi \in \Gamma(\Sigma, \mathbf{g})$ is a zero form on Σ with value in the adjoint representation. It is often called a boson. ϵ is an anti-commuting parameter. We can also write

$$\delta\Phi = iD\Phi = -i\{Q, \Phi\}$$

for every field Φ. Here $Q = -D$ is the analog of twisted differential. We have $Q^2 = -i\delta_\phi$. So for gauge equivalent field, we have $Q^2 = 0$.

Let V be a gauge invariant functional. Let $L = -i\{Q, V\}$. The following integral defines a cohomolgical field theory:

$$Z = \int \mathcal{D}\Phi e^{-i\{Q,V\}}.$$

The space of connections is an affine space with a natural symplectic form. The gauge group acts on the space of connections which preserves the symplectic form. The moment map of the action is the Yang-Mills functional. So this falls into the case of localization of symplectic action and the path integral is then reduced to an integral over the fixed point, especially the fixed point set of absolute minimals for polynomial expressions. So the above integral localize at fixed points of the group action, and we have

$$Z = \sum_{\sigma \in S} Z_\sigma.$$

Z_σ denotes contributions over a component of the fixed point set X_σ, it also depends on the normal bundle of the component. An important point is that a properly chosen cohomological theory is equivalent to the physical Yang-Mills theory.

6.4.2 *Relation with physical Yang-Mills theory*

The physical Yang-Mills theory is given by

$$Z = \frac{1}{\mathrm{vol}(\mathcal{G})} \int \mathcal{D}Ae^{-L},$$

where $L = \frac{1}{4e^2}\int d^2x|*F_A|^2$ and $*$ is the star operator with respect to a chosen metric.

If we introduce a scalar field ϕ with values in g, then the above integral can be written as

$$Z = \frac{1}{\mathrm{vol}(\mathcal{G})}\int \mathcal{D}A\mathcal{D}\phi \exp\{i\int \mathrm{Tr}\phi F - \frac{e^2}{2}\int d\mu \mathrm{Tr}\phi^2\}.$$

$d\mu = *(1)$ is a measure with respect to a chosen metric. One can also add fermion field $\phi \in \Gamma(\Sigma, \Omega^1(\Sigma)\otimes \mathrm{g})$ which lies in the tangent space to the space of connections. The path integral is the same as

$$Z = \frac{1}{\mathrm{vol}(\mathcal{G})}\int \mathcal{D}A\mathcal{D}\psi\mathcal{D}\phi \exp\{i\int_\Sigma \mathrm{Tr}(\phi F) + \frac{1}{2}\mathrm{Tr}(\psi\wedge\psi) - \frac{e^2}{2}\int d\mu \mathrm{Tr}(\phi^2)\}.$$

It is easy to check that the Lagrangian is gauge invariant. The derivation from cohomological Yang-Mills theory to the physical Yang-Mills theory can be found on pp 32-40 of Witten's paper [Witten-localization]. In particular, the following precise correspondence between them is established:

$$< \exp(\omega + \epsilon\Theta)\beta >$$

$$= \frac{1}{\mathrm{vol}(\mathcal{G})}\int \mathcal{D}A\mathcal{D}\psi\mathcal{D}\phi \exp(\frac{1}{4\pi^2}\int_\Sigma \mathrm{Tr}(i\phi F + \frac{1}{2}\psi\wedge\psi) + \frac{\epsilon}{8\pi^2}\int_\Sigma d\mu \mathrm{Tr}\phi^2)\beta.$$

The left-hand side is the cohomological Yang-Mills theory. With the help of localization the original path integral is reduced to the integral over the space of flat connections $\mathcal{M}$. Here $\omega = \frac{1}{4\pi^2}\int_\Sigma \mathrm{Tr}(i\phi F + \frac{1}{2}\psi\wedge\psi)$ is the natural symplectic form on $\mathcal{M}$. Note that $T^*\mathcal{M} = H^1(\Sigma, T\Sigma\otimes \mathrm{g})$. The definition of ω only depends on the cohomological class of $i\phi F + \frac{1}{2}\psi\wedge\psi.\Theta = \frac{1}{8\pi^2}\int_\Sigma d\mu \mathrm{Tr}\phi^2$ is also a well-defined form on $\mathcal{M}.\omega, \Theta$ are the fundamental BRST invariant observables.

There is one more BRST observable $V_C = \frac{1}{4\pi^2}\int_C \mathrm{Tr}\phi\psi$. Here $C \subset \Sigma$ is a simple closed curve. In general, one should evaluate

$$\int_{\mathcal{M}} \exp(\omega + \epsilon\Theta + \Sigma_\sigma \eta_\sigma V_\sigma).$$

From the physical Yang-Mills theory, after we perform Gaussian integral on ψ, we eliminate the V_C terms. This is why we have the above integral which is the same as

$$\int_{\mathcal{M}} \exp(\omega + \bar{\epsilon}\Theta)$$

where $\bar{\epsilon} = \epsilon - 2\sum_{\sigma<\tau} \eta_\sigma \eta_\tau \gamma_{\sigma\tau}, \gamma_{\sigma\tau} = \sharp(C_\sigma \cap C_\tau)$. Again $\mathcal{M}$ is the space of flat connections on Σ which are absolute minimals of Yang-Mills functionals.

6.4.3 *Evaluation of Yang-Mills theory*

We now turn to more concrete calculations of quantizations of two dimensional Yang-Mills theory.

First we consider the case of cylinder $C \times [0, T]$, where $C \simeq S^1$ is a circle. The space of fields is equal to the space of connections A_C on the circle. So the Hilbert space is $\mathcal{H} = C^\infty(A_C)^{\mathcal{G}}$, the space of gauge invariant functions on A_C.

Since up to gauge transformation all connections are classified by their holonomy around the curve C, we can identify the Hilbert space with $\mathcal{H} = C^\infty(G)^G$, i.e. the ring of functions on G which are invariant under the adjoint action, or equivalently functions on the space of conjugacy classes which is the same as functions on the maximal torus modulo the action of the Weyl group. This is nothing but the characters of irreducible representations of G. We take characters χ_R to form a basis of $\mathcal{H}, \mathcal{H} = \oplus_R \mathbf{C}_R$. The state associated with C and A is then $\Psi_R(A) = \chi_R(\mathrm{Hol}_C A)$, i.e. the character of the holonomy of A along C with respect to a representation R.

In canonical quantization, $A^0(\theta)d\theta$ and $-i\frac{\partial}{\partial A^0(\theta)}$ are conjugating variables. So the Hamiltonian is $H = \frac{e^2}{2}\int_C \mathrm{Tr}(\frac{\delta}{\delta A})^2$. Here $\frac{\delta}{\delta A^0(\theta)}$ acts on Ψ_R by $\frac{\delta}{\delta A^0(\theta)}\mathrm{Tr}_R \mathrm{Hol}_C(A) = \mathrm{Tr}_R T^a \mathrm{Hol}_C(A)$.

We see that,

$$\sum_a \frac{\delta}{\delta A^a}\frac{\delta}{\delta A^a}\mathrm{Tr}_R \mathrm{Hol}_C(A) = \mathrm{Tr}_R(\sum_a T^a T^a \mathrm{Hol}_C(A)),$$

$\sum_a T^a T^a$ is the quadratic Casimir of $G, \mathrm{Tr}_R \sum_a T^a T^a = c_2(R)$. Thus, we have

$$H\Psi_R = \frac{e^2 c_2(R) L}{2}\Psi_R,$$

$$< \Psi_R | e^{-HT} | \Psi_R >= \exp(-e^2 a \frac{c_2(R)}{2}),$$

$$Z_R = \omega_R \exp(-c_2(R)\frac{ae^2}{2}).$$

Next, we consider a sphere with three boundaries C_1, C_2, C_3. Naively the partition function is $Z = \sum_{R_1,R_2,R_3} \omega_{R_1,R_2,R_3}$. By the consideration of Casimir which is the same for all invariant polynomials one gets that $R_1 = R_2 = R_3$. Hence the partition function is: $Z = \sum_R \omega_R \Pi_{\gamma=1}^3 \Psi_R(A|_{C_\gamma})$.

For a closed surface Σ of genus g, we cut the surface into $2g-2$ disjoint copies of three punctured spheres along $3g-3$ disjoint simple closed curves. By using the orthogonality relation of group characters which correspond to states R, we have

$$Z(\Sigma) = \sum_R \omega_R^{2g-2} \exp(-\frac{e^2 a c_2(R)}{2}).$$

To compute ω_R, we first calculate Z on a disk D, $Z(D) = v_R \exp(-\frac{e^2 a c_2(R)}{2})$. Consider:

$$Z(S^1 \times I) = \exp(-\frac{e^2 a c_2(R)}{2}),$$

$$Z(P) = \omega_R \Pi_{\gamma=1}^3 \psi_R(A|_{C_\gamma}),$$

$$Z(D) = v_R \exp(-\frac{e^2 a c_2(R)}{2}),$$

$$Z(S^1 \times I) = Z(\Sigma_1) \times Z(D).$$

This implies that $v_R w_R = 1$. On D, we fix the holonomy of a connection around ∂D to take value in $U \in G$. And $Z(U) = \sum_R v_R \chi_R(U)$. On the other hand, we have $Z(U) = \exp \alpha\delta(U-1)$. By using orthogonality relations for group characters, we find $v_R = \exp(\alpha) \dim R$. Hence,

$$Z(\Sigma^g, e^2 a) = \sum_R \omega_R^{2g-2} \exp(-\frac{c_2(R) e^2 a}{2}) = \Sigma \frac{e^{-\alpha(2g-2)} \exp(-c_2(R)\frac{\epsilon}{2})}{(\dim R)^{2g-2}}.$$

Now, consider an example $G = SU(2)$. For each integer $n \geq 1$, there is a unique representation R_n of dimension n. The quadratic Casimir of representation R_n is $c_2(R_n) = \frac{n^2-1}{2}$. We normalize it to $c_2(\bar{R}_n) = \frac{n^2}{2}$. Hence the partition function is:

$$Z(\Sigma, \epsilon) = e^{-\alpha(2g-2)} \sum_{n \geq 1} \frac{e^{-\epsilon n^2/4}}{n^{2g-2}}.$$

We will show that Z can be written as a sum over critical points to compare it with the formula by localization principle. We start with

$$\frac{\partial^{g-1}}{\partial \epsilon^{g-1}} Z(\Sigma, \epsilon) = \frac{e^{-\alpha(2g-2)}}{2^{2g-1}} (-1 + \sum_{n \in \mathbf{Z}} \exp(-\frac{\epsilon n^2}{4}).$$

By using Poisson summation formula we have

$$\frac{\partial^{g-1}}{\partial\epsilon^{g-1}}Z = \frac{e^{-\alpha(2g-2)}}{2^{2g-1}}(-1+\sqrt{\frac{4\pi}{\epsilon}}\sum_{m\in\mathbf{Z}}\exp(-\frac{(2\pi m)^2}{\epsilon})).$$

This agrees with the formula by localization principle from the following consideration. The Eular-Lagrangian equation of Yang-Mills is $Df = 0$. If $f \neq 0$, being covariantly constant, this gives a reduction of the structure group of the connection to a subgroup H_0 that commutes with f. Solutions are therefore flat H_0 connections twisted by constant curvature line bundles in the $U(1)$ subgroup generated by f. In the case of $G = SU(2)$, we get an $SU(2)$ bundle with a covariantly constant splitting as a sum of line bundles. From the classification of line bundles, it follows that the conjugacy class of f is given by

$$f = 2\pi m \begin{pmatrix} i & 0 \\ 0 & -i \end{pmatrix}$$

with $m \in \mathbf{Z}$. The value of I at such a critical point is $I_m = \frac{(2\pi m)^2}{\epsilon'}$. This shows that it agrees with the formula by localization principle.

6.5 Combinatorial approach to 2D Yang-Mills theory

Let Σ be a two dimensional surface, $E \to \Sigma$ be a G bundle with G a Lie group. Let A be a connection on Σ, i.e. a $\mathcal{G}$ valued one form. As usual, $F = dA + 1/2 A \wedge A$ be its curvature. F is $\mathcal{G}$ valued two form. Pick up a metric on Σ, it induces a Hodge $*$ operator on forms. Since Σ is two dimensional, $f = *F = \epsilon F$ is a $\mathcal{G}$ valued zero form, here ϵ is the area form. As usual the Yang-Mills Lagrangian is defined as:

$$I(A) = -\frac{1}{2e^2}\int_\Sigma Tr(F \wedge *F) = -\frac{1}{2e^2}\int_\Sigma Tr(f^2)$$

$I(A)$ actually only depends on A and the measure $d\mu$ induced by the metric. Let $\rho = \int_\Sigma d\mu$. It is useful to notice that $I(A)$ is invariant under $e^2 \to te^2, \mu \to \mu/t$.

The quantization of Yang-Mills in two dimensions is to compute the partition function:

$$Z_\Sigma(e^2,\rho) = \int \mathcal{D}Ae^{-I(A)}.$$

An alternative Lagrangian is

$$I'(A,\phi) = -\frac{e^2}{2}\int_\Sigma d\mu Tr\phi^2 - i\int_\Sigma Tr\phi F.$$

Their partition function

$$\bar{Z}_\Sigma(e^2\rho) = \int \mathcal{D}\phi\mathcal{D}Ae^{\frac{e^2}{2}\int d\mu Tr\phi^2 + i\int Tr\phi F}$$

is the same as two dimensional Yang-Mills because one can integrate ϕ the same way as in one dimension:

$$\int_{-\infty}^{\infty}\frac{dx}{\sqrt{2\pi}}e^{-\frac{e^2}{2}x^2 - ixy} = e^{-\frac{y^2}{2e^2}}.$$

Let $e \to 0$, one has a topological field theory $Z_\Sigma(0)$. We want to solve this model.

To do so, we triangulate Σ into plaquettes w_i, i.e. polygons with boundary $U_1, U_2, ..., U_n$. Each w_i carries a measure $\rho_i, \Sigma\rho_i = \rho$. We have

$$e^{-\int_\Sigma \mathcal{I}} = \Pi_i e^{-\int_{w_i}\mathcal{I}}.$$

Here is a key observation: The conjugacy class of the holonomy is gauge invariant, the local factor associated with a plaquette must be a class function of the holonomy. Class function must be a linear combination of the group characters which forms a basis of class functions.

For each plaquette and A we have holonomy $\mathcal{U} = \int_{\partial w} A$. Let α be an irreducible representation of G, we define $\chi_\alpha(\mathcal{U}) = Tr_\alpha\mathcal{U}$. It is Midgal who first made an Ansatz:

$$\Gamma(\mathcal{U},\rho) = \Sigma_\alpha \dim\alpha\chi_\alpha(\mathcal{U})e^{(-\frac{\rho_w c_2(\alpha)}{2})}.$$

And

$$Z_\Sigma(\rho) = \int \Pi_\gamma dU_\gamma \Pi_i \Gamma(\mathcal{U}_i, \rho_i).$$

The crucial propety is that it is invariant under subdivision. The proof is as follows.

Let

$$\Gamma = \Sigma_\alpha \dim \alpha \chi_\alpha(U_1U_2U_3U_4)e^{(-\frac{\rho_0 c_2(\alpha)}{2})}.$$

be for the plaquette $U_1U_2U_3U_4$. We add a line V so it is divided into two small plquettes U_1U_2V and $V^{-1}U_3U_4$. We have

$$\Gamma'\Gamma'' = \Sigma_{\alpha,\beta} \dim \alpha \dim \beta \chi_\alpha(U_1U_2V)\chi_{|beta}(V^{-1}U_3U_4)e^{(-\frac{\rho' c_2(\alpha)}{2})}e^{(-\frac{\rho'' c_2(\beta)}{2})}$$

We have orthogonal relation for characters:

$$\int dV \chi_\alpha(AV)\chi_\beta(V^{-1}B) = \delta_{\alpha\beta}\frac{1}{\dim \alpha}\chi_\alpha(AB).$$

The above implies that

$$\int dV \Gamma'\Gamma'' = \Gamma.$$

This gives invariance under subdivision.

As an application, we represent a close surface Σ as a disk with boundary $a_1b_1a_1^{-1}b_1^{-1}a_2b_2a_2^{-1}b_2^{-1}...$, then we have the partition function for Σ as:

$$Z_\Sigma(\rho) = \Sigma_\alpha \dim \alpha e^{-\frac{\rho c_2(\alpha)}{2}} \int dU_i dV_j \chi_\alpha(U_1V_1U_1^{-1}V_1^{-1}...U_gV_gU_g^{-1}V_g^{-1})$$

$$= \Sigma_\alpha \frac{e^{-\frac{-\rho c_2(\alpha)}{2}}}{(\dim \alpha)^{2g-2}}.$$

Remark: The above partition function can be identified as the symplectic volume.

Remark: A rigorous mathematical justification based on above localization for Yang-Mills is available by works of Liu Kefeng [Liu1], [Liu2] and Jeffery and Kirwan [Jeffery-Kirwan].

Reprinted from *Complex Manifold Without Potential Theory* by S. S. Chern, Revised edition, Springer, 1995, pp. 148–154.

7. Chern-Simons Invariant of Three-dimensional Manifolds

In (125), §4 we defined an invariant J = I(s) mod 1, for a compact oriented three-dimensional manifold. This has been called a Chern-Simons invariant. It has played an important role in both mathematics and physics. In fact, up to an additive constant it is the eta invariant of the manifold, as introduced by Atiyah, Patodi, and Singer via spectral theory [1]; cf. Ref to §7. It has also been used by W. Thurston in his theory of hyperbolic manifolds, while Robert Meyerhoff has shown that, for certain hyperbolic manifolds, it takes values which are dense on the unit circle. In mathematical physics the concept is found useful in quantum field theory, statistical mechanics, and the theory of anyons.

We begin by repeating its definition. Let M denote the manifold, oriented, and let P be the bundle of its orthonormal frames, so that we have

(198) $$\pi : P \to M,$$

where π is the projection, mapping a frame $xe_1e_2e_3$ to its origin x. A section $s : M \to P$ of the bundle satisfies the condition $\pi \circ s$ = identity and can be viewed as a field of frames. It is well known that in our case such a section always exists; we say that M is *parallelizable*.

To such a frame field the *Levi-Civita connection* of the metric is given by an antisymmetric matrix of one-forms:

(199) $$\varphi = (\varphi_{ij}), \qquad 1 \leq i,\ j \leq 3,$$

and its *curvature* by an antisymmetric matrix of two-forms:

(200) $$\Phi = (\Phi_{ij}), \qquad 1 \leq i,\ j \leq 3.$$

Throughout this section our small Latin indices will run from 1 to 3. We have

$$d\varphi_{ik} = \sum_j \varphi_{ij} \wedge \varphi_{jk} + \Phi_{ik}. \tag{201}$$

We introduce the three-form

$$T = \frac{1}{8\pi^2} \sum_{i,j,k} \left(\varphi_{ij} \wedge \Phi_{ij} - \tfrac{1}{3} \varphi_{ij} \wedge \varphi_{jk} \wedge \varphi_{ki} \right), \tag{202}$$

and consider the integral

$$\Phi(s) = \int_M \tfrac{1}{2} T. \tag{203}$$

It will be proved below that for another section $t : M \to P$, $\Phi(t) - \Phi(s)$ is an integer, so that $\Phi(s)$ mod 1 is independent of s. This defines an invariant $J(M) \in \mathbb{R}/\mathbb{Z}$, where $J(M) = \Phi(s)$ mod 1.

In [3] we proved the theorems:

<u>Theorem 7.1.</u> *J(M) is a conformal invariant, i.e. it remains unchanged under a conformal transformation of the metric.*

<u>Theorem 7.2.</u> *J(M) has a critical value at M if and only if M is locally conformally flat.*

We wish to give direct proofs of these theorems in this section.

<u>A) Family of Connections</u>

We shall use arbitrary frame fields to develop the Riemannian geometry on M. Let $xe_1e_2e_3$ be a frame, and ω^1, ω^2, ω^3, its dual coframe. Let the inner product be

$$\langle e_i, e_j \rangle = g_{ij}. \tag{204}$$

We introduce g^{ij} through the equations

$$\sum g_{ij} g^{jk} = \delta_i^k \tag{205}$$

and use the g's to raise or lower indices, as in classical tensor analysis. Then the *connection forms* ω_i^j or ω_{ij} are determined, uniquely, by the equations

$$d\omega^i = \sum \omega^j \wedge \omega_j^i, \qquad \omega_{ij} + \omega_{ji} = dg_{ij}. \tag{206}$$

The *curvature forms* are defined by

$$\Omega_i^j = d\omega_i^j - \sum \omega_i^k \wedge \omega_k^j. \tag{207}$$

By exterior differentiation of (207) we get the *Bianchi identity*

$$d\Omega_i^j = \sum \omega_i^k \wedge \Omega_k^j - \sum \Omega_i^k \wedge \omega_k^j. \tag{208}$$

To avoid confusion notice our convention that the upper index in ω_i^j, Ω_i^j is the second index. Thus $\sum_j \omega_i^j \, g_{jk} = \omega_{ik}$ ($\neq \omega_{ki}$ in general).

We introduce the cubic form

$$8\pi^2 T = -\frac{1}{3}\sum \omega_i^j \wedge \omega_j^k \wedge \omega_k^i + \sum \omega_i^i \wedge \Omega_j^j - \sum \omega_i^j \wedge \Omega_j^i. \tag{209}$$

On our three-manifold M, T is of course closed. But the basic reason for its importance and interesting properties is that formally by (207), (208), we have

$$\begin{aligned} 8\pi^2 dT &= \sum \left(\Omega_i^i \wedge \Omega_j^j - \Omega_i^j \wedge \Omega_j^i\right) \\ &= \sum \delta_{i_1 i_2}^{j_1 j_2} \Omega_{j_1}^{i_1} \wedge \Omega_{j_2}^{i_2}, \end{aligned} \tag{210}$$

which is the *first Pontrjagin form.*

Consider a family of connections on M, depending on a parameter τ. Then ω_i^j, Ω_i^j, T all involve τ, and we have the fundamental formula

$$\begin{aligned} 8\pi^2 \frac{\partial T}{\partial \tau} &= -d\left\{\sum \omega_i^i \wedge \frac{\partial \omega_j^j}{\partial \tau} - \sum \omega_i^j \frac{\partial \omega_j^i}{\partial \tau}\right\} \\ &\quad + 2\sum \left(\frac{\partial \omega_i^i}{\partial \tau} \wedge \Omega_j^j - \frac{\partial \omega_i^j}{\partial \tau} \wedge \Omega_j^i\right). \end{aligned} \tag{211}$$

The proof of (211) is straightforward. It follows by differentiation of (209), and using the formulas obtained by differentiation of (206), (207) with respect to τ. It is useful to observe that the last term is a polarization of the Pontrjagin form.

Let P' be the bundle of all frames of M, so that we have

$$P \xrightarrow{i} P' \qquad \pi \searrow \swarrow \pi' \qquad M \tag{212}$$

where i is the inclusion. Then T in (209) can be considered as a form in P' and its pull-back $i*T$ is the T given by (202). We will make such identifications when there is no danger of confusion.

We consider $\Phi(s)$ defined by (203). When t is another section, then $t(M) - s(M)$, as a three-dimensional cycle is homologous, modulo torsion, to an integral multiple of the fiber P_x, $x \in M$. But P_x is topologically SO(3) and ω_{ij} reduces on it to its Maurer-Cartan forms. If $j : P_x = SO(3) \to P$ is the inclusion,

$$\pm\frac{1}{2}\, j * T = \frac{1}{8\pi^2}\omega_{12} \wedge \omega_{23} \wedge \omega_{31},$$

whose integral over P_x is 1. Hence $\Phi(s)$ mod 1 is independent of s.

We wish to clarify the relation between orthonormal frame fields and arbitrary frame fields. By the Schmidt orthogonalization process P is a retract of P', under which the origin of the frame is fixed. The retraction we denote by $r : P' \to P$. We consider the form T defined in (209) to be in P'. If $s' : M \to P'$ is a section, then $s = r \circ s'$ is also a section and they are homotopic through sections. Since $dT = 0$, we have

$$\int_{s'M} T = \int_{sM} i*T. \tag{213}$$

Hence J(M) can be computed through an arbitrary frame field by the left-hand side of the last equation.

B) Proofs of the Theorems

In view of the above remark we can, for local considerations, use a local coordinate system u^i and the resulting natural frame field $\partial/\partial u^i$. We shall summarize the well-known formulas, which are

$$\omega_i^j = \sum \Gamma_{ik}^j du^k,$$

(214)
$$\Gamma_{ik}^j = \sum g^{jl}\Gamma_{ilk},$$

$$\Gamma_{ilk} = \frac{1}{2}\left(\frac{\partial g_{il}}{\partial u^k} + \frac{\partial g_{kl}}{\partial u^i} - \frac{\partial g_{ik}}{\partial u^l}\right),$$

$$\Omega_i^j = \frac{1}{2}\sum R_{ikl}^j du^k \wedge du^l.$$

The R_{ijkl} satisfy the symmetry relations

(215)
$$R_{ijkl} = -R_{jikl} = -R_{ijlk}, \quad R_{ijkl} = R_{klij},$$

$$R_{ijkl} + R_{iklj} + R_{iljk} = 0, \quad R_{ijkl,h} + R_{ijlh,k} + R_{ijhk,l} = 0,$$

where the comma denotes covariant differentiation. They imply

(216)
$$\Omega_{ij} + \Omega_{ji} = 0, \quad \sum \Omega_{ij} \wedge dx^j = 0,$$

and

(217)
$$\sum \Omega_i^i = \sum g^{ik}\Omega_{ik} = 0.$$

We also introduce the *Ricci curvature* and the *scalar curvature by*

(218)
$$R_k^i = \sum_j R_{kj}^{ij}, \quad R = \sum R_i^i.$$

For treatment of the conformal geometry we define

(219)
$$C_{kl}^j = R_{k,l}^j - R_{l,k}^j - \frac{1}{4}(\delta_k^j R_{,l} - \delta_l^j R_{,k}),$$

$$C_{jkl} = R_{jk,l} - R_{jl,k} - \frac{1}{4}(g_{jk}R_{,l} - g_{jl}R_{,k}).$$

Then the Bianchi identities give

(220)
$$\sum C_{kj}^j = 0, \quad C_{jkl} + C_{klj} + C_{ljk} = 0.$$

These relations imply that the matrix

$$(221)\qquad C = \begin{bmatrix} C^1_{23} & C^1_{31} & C^1_{12} \\ C^2_{23} & C^2_{31} & C^2_{12} \\ C^3_{23} & C^3_{31} & C^3_{12} \end{bmatrix}$$

is symmetric and that the matrix GC, where $G = (g_{ij})$, has trace zero.

Schouten proved [4, p. 92] that the three-dimensional manifold M is conformally flat, if and only if $C = 0$, i.e. $C_{ijk} = 0$.

By integrating (211), we get

$$(222)\qquad 8\pi^2 \frac{\partial}{\partial \tau} \int_M T = 2 \int_M \Delta,$$

where

$$(223)\qquad \Delta = \sum \left(\frac{\partial \omega^i_i}{\partial \tau} \wedge \Omega^j_j - \frac{\partial \omega^j_i}{\partial \tau} \wedge \Omega^i_j \right) = - \sum \frac{\partial \omega^j_i}{\partial \tau} \wedge \Omega^i_j,$$

by (217).

We consider a family of metrics $g_{ij}(\tau)$ and put

$$(224)\qquad v_{ij} = \frac{\partial g_{ij}}{\partial \tau}.$$

To prove Theorem 7.1 we suppose this is a conformal family of metrics, i.e.

$$(225)\qquad v_{ij} = \sigma g_{ij}.$$

From (214) we find

$$(226)\qquad \frac{\partial \omega^j_i}{\partial \tau} = \frac{1}{2} \sum (\delta^j_i \sigma_k + \delta^j_k \sigma_i - g_{ik} g^{jl} \sigma_l) dx^k,$$

where $\sigma_k = \partial\sigma/\partial x^k$. By the second equation of (216) and the equation (217) we find $\Delta = 0$. This proves that $\int_M T$ is independent of τ, and hence Theorem 7.1.

To prove Theorem 7.2 we consider v_{ij} such that the trace

$\sum v_i^i = 0$. Geometrically this means that we consider the tangent space of the space of conformal structures on M. From (214) we find

(227)

$$\frac{\partial \omega_i^j}{\partial \tau} = \sum g^{jl} \left\{ -\sum v_{lk}\omega_i^k + \frac{1}{2}\sum \left(\frac{\partial v_{il}}{\partial u^k} + \frac{\partial v_{kl}}{\partial u^i} - \frac{\partial v_{ik}}{\partial u^l} \right) du^k \right\}.$$

It follows that

(228)

$$\Delta = \sum_{i,j} \Omega^{ij} \left\{ -\sum v_{jk}\omega_i^k + \frac{1}{2} dv_{ij} + \frac{1}{2}\sum_k \left(\frac{\partial v_{kj}}{\partial u^i} - \frac{\partial v_{ki}}{\partial u^j} \right) du^k \right\}.$$

The term in the middle is zero, because Ω^{ij} is antisymmetric and dv_{ij} is symmetric in i, j. To the integral of the last term we apply Stokes theorem to reduce it to an integral involving only the v_{ij}, and not their derivatives. We will omit the details of this computation. The result is that the condition

$$\frac{\partial}{\partial \tau} \int_M T = 0$$

is equivalent to

(229) $$\int_M \mathrm{Tr}(VC)\; du^1 \wedge du^2 \wedge du^3 = 0.$$

If the metric is conformally flat, we have C = 0 and hence the vanishing of the above integral.

Conversely, at a critical point of Φ we must have Tr(VC) = 0 for all symmetric V satisfying $\mathrm{Tr}(VG^{-1}) = 0$. Hence C is a multiple of G^{-1} or GC is a multiple of the unit matrix. But GC has trace zero. Hence it must itself be zero and we have C = 0. This proves Theorem 7.2.

J. DIFFERENTIAL GEOMETRY
33 (1991) 787–902

GEOMETRIC QUANTIZATION OF CHERN-SIMONS GAUGE THEORY

SCOTT AXELROD, STEVE DELLA PIETRA & EDWARD WITTEN

Abstract

We present a new construction of the quantum Hilbert space of Chern-Simons gauge theory using methods which are natural from the three-dimensional point of view. To show that the quantum Hilbert space associated to a Riemann surface Σ is independent of the choice of complex structure on Σ, we construct a natural projectively flat connection on the quantum Hilbert bundle over Teichmüller space. This connection has been previously constructed in the context of two-dimensional conformal field theory where it is interpreted as the stress energy tensor. Our construction thus gives a $(2+1)$-dimensional derivation of the basic properties of $(1+1)$-dimensional current algebra. To construct the connection we show generally that for affine symplectic quotients the natural projectively flat connection on the quantum Hilbert bundle may be expressed purely in terms of the intrinsic Kähler geometry of the quotient and the Quillen connection on a certain determinant line bundle. The proof of most of the properties of the connection we construct follows surprisingly simply from the index theorem identities for the curvature of the Quillen connection. As an example, we treat the case when Σ has genus one explicitly. We also make some preliminary comments concerning the Hilbert space structure.

Introduction

Several years ago, in examining the proof of a rather surprising result about von Neumann algebras, V. F. R. Jones [20] was led to the discovery of some unusual representations of the braid group from which invariants of links in $\mathbf{S}^3$ can be constructed. The resulting "Jones polynomial" of links has proved in subsequent work to have quite a few generalizations, and to be related to two-dimensional lattice statistical mechanics and to quantum groups, among other things.

Received by the editors November 6, 1989, and, in revised form, January 23, 1990. The first author's research was supported in part by a National Science Foundation Graduate Fellowship and the Alfred P. Sloan Foundation. The second author's research was supported in part by National Science Foundation Grant 86-20266 and National Science Foundation Waterman Grant 88-17521.

Tsuchiya and Kanie [38] recognized that the Jones braid representations and their generalizations coincide with certain representations of braid groups and mapping class groups that have quite independent origins in conformal field theory [4] and that have been intensively studied by physicists [11], [40], [23], [31]. (The representations in question are actually projective representations, for reasons that will be clear later.) The conformal field theory viewpoint leads to a rigorous construction of these representations [34], [39].

Conformal field theory alone, however, does not explain why these particular representations of braid groups and mapping class groups are related to *three-dimensional* invariants. It was conjectured [2] that some form of three- or four-dimensional gauge theory would be the key to understanding the three-dimensional invariances of the particular braid traces that lead to the Jones polynomial. Recently it has been shown [41] that three-dimensional Chern-Simons gauge theory for a compact gauge group G indeed leads to a natural framework for understanding these phenomena. This involves a nonabelian generalization of old work by A. Schwarz relating analytic torsion to the partition functions of certain quantum field theories with quadratic actions [32], and indeed Schwarz had conjectured [33] that the Jones polynomial was related to Chern-Simons gauge theory.

Most of the striking insights that come from Chern-Simons gauge theory depend on use of the Feynman path integral. To make the path integral rigorous would appear out of reach at present. Of course, results predicted by the path integral can be checked by, e.g., showing that the claimed three-manifold invariants transform correctly under surgery, a program that has been initiated in [30]. Such combinatorial methods—similar to methods used in the original proofs of topological invariance of the Jones polynomial—give a verification but not a natural explanation of the three-dimensional symmetry of the constructions.

In this paper, we pursue the more modest goal of putting the Hamiltonian quantization of Chern-Simons gauge theory—which has been discussed heuristically in [42] and in [7]—on a rigorous basis. In this way we will obtain new insights about the representations of braid and mapping class groups that arise in this theory. These representations have been constructed, as we have noted, from other points of view, and most notably from the point of view of conformal field theory. However, three-dimensional quantum field theory offers a different perspective, in which the starting point is the fact that affine spaces and their symplectic quotients can be quantized in a natural way. Our goal in this paper is to give a rigorous construction of the representations of mapping class groups that are associated with the Jones polynomial, from the point of view of the three-dimensional quantum field theory.

Canonical quantization. The goal is to associate a Hilbert space to every closed oriented 2-manifold Σ by canonical quantization of the Chern-Simons theory on $\Sigma \times R$. As a first step, we construct the physical phase space, $\mathscr{M}$. It is the symplectic quotient of the space, $\mathscr{A}$, of G connections on Σ by the group $\mathscr{G}$ of bundle automorphisms. It has a symplectic form ω which is k times the most fundamental quantizable symplectic form ω_0; here k is any positive integer. $\mathscr{M}$ is the finite-dimensional moduli space of flat G connections on Σ. We then proceed to quantize $\mathscr{M}$ as canonically as possible. We pick a complex structure J on Σ. This naturally induces a complex structure on $\mathscr{M}$, making it into a Kähler manifold. We may then construct the Hilbert space $\mathscr{H}_J(\Sigma)$ by Kähler quantization. If $\mathscr{T}$ denotes the space of all complex structures on $\mathscr{M}$, we thus have a bundle of Hilbert spaces $\mathscr{H}(\Sigma) \rightarrow \mathscr{T}$. This "quantum bundle" will be denoted $\tilde{\mathscr{H}}_Q$. For our quantization to be "canonical" it should be independent of J, at least up to a projective factor. This is shown by finding a natural projectively flat connection on the quantum bundle.

The essential relation between Chern-Simons gauge theory and conformal field theory is that this projectively flat bundle is the same as the bundle of "conformal blocks" which arises in the conformal field theory of current algebra for the group G at level k. This bundle together with its projectively flat connection is relatively well understood from the point of view of conformal field theory. (In particular, the property of "duality" which describes the behavior of the $\mathscr{H}_J(\Sigma)$ when Σ degenerates to the boundary of moduli space has a clear physical origin in conformal field theory [4], [40]. The property is essential to the computability of the Jones polynomial.) The conformal field theory point of view on the subject has been developed rigorously from the point of view of loop groups by Segal [34], and from an algebra-geometric point of view by Tsuchiya et al. [39]. Also, there is another rigorous approach to the quantization of $\mathscr{M}$ due to Hitchin [18]. Finally, in his work on non-abelian theta functions, Fay [8] (using methods more or less close to arguments used in the conformal field theory literature) has described a heat equation obeyed by the determinant of the Dirac operator which is closely related to the construction of the connection and may in fact lead to an independent construction of it.

We will be presenting an alternative description of the connection on $\mathscr{H}(\Sigma)$ which arises quite naturally from the theory of geometric quantization. In fact, this entire paper is the result of combining three simple facts.

(1) The desired connection and all of its properties are easily understood for Kähler quantization of a finite-dimensional affine symplectic manifold $\mathcal{A}$. In that case the connection 1-form is a simple second order differential operator on $\mathcal{A}$ acting on vectors in the quantum Hilbert space (which are sections of a line bundle over $\mathcal{A}$).

(2) By geometric invariant theory we can present a simple abstract argument to "push down" this connection "upstairs" for quantization of $\mathcal{A}$ to a connection "downstairs" for the quantization of $\mathcal{M}$. Here, $\mathcal{M}$ is the symplectic quotient of $\mathcal{A}$ by a suitable group of affine symplectic transformations that preserves the complex structure which is used in quantizing $\mathcal{A}$.[1] The one-form $\mathcal{O}$ for the connection downstairs is a second-order differential operator on $\mathcal{M}$.

(3) Even in the gauge theory case where the constructions upstairs are not well-defined since $\mathcal{A}$ is infinite dimensional, we may present the connection downstairs in a well-defined way. We first work in the finite dimensional case and write out an explicit description of $\mathcal{O}$. We then interpret the "downstairs" formulas in the gauge theory case in which the underlying affine space is infinite dimensional though its symplectic quotient is finite dimensional. As is familiar from quantum field theory, interpreting the "downstairs" formulas in the gauge theory context requires regularization of some infinite sums. This can, however, be done satisfactorily.

What has just been sketched is a very general strategy. It turns out that we have some "luck"—the definition of $\mathcal{O}$ and the proof of most of its properties can all be written in terms of the Kähler structure of $\mathcal{M}$ and a certain regularized determinant which is independent of the quantization machinery. Since these objects refer only to $\mathcal{M}$, our final results are independent of geometric invariant theory. One consequence of this independence is that our results apply for an arbitrary prequantization line bundle on $\mathcal{M}$, and not just for line bundles which arise as pushdowns of prequantum line bundles on $\mathcal{A}$.

The infinite dimensionality of the affine space that we are studying shows up at one key point. Because of what physicists would call an "anomaly", one requires a rescaling of the connection 1-form from the normalization it would have in finite dimensions. This has its counterpart in conformal field theory as the normalization of the Sugawara construction [14], which is the basic construction giving rise to the connection from that point of view. This rescaling does not affect the rest of the calculation

[1] For physicists, geometric invariant theory is just the statement that, in this situation, one gets the same result by imposing the constraints corresponding to $\mathcal{G}$ invariance before or after quantization.

except to rescale the final answer for the central curvature of the connection. This reproduces a result in conformal field theory. From our point of view, though we can describe what aspects of the geometry of the moduli space lead to the need to rescale the connection, the deeper meaning of this step is somewhat mysterious.

Outline. This paper is quite long. Essentially this is because we rederive the connection several times from somewhat different viewpoints and because we describe the special case of genus one in considerable detail. Most readers, depending on their interests, will be able to omit some sections of the paper.

For physicists, the main results of interest are mostly in §§2 and 5, and amount to a $(2+1)$-dimensional derivation of the basic properties of $(1+1)$-dimensional current algebra, including the values of the central charge and the conformal dimensions. This reverses the logic of previous treatments in which the understanding of the $(2+1)$-dimensional theory ultimately rested, at crucial points, on borrowing known results in $(1+1)$-dimensions. This self-contained $(2+1)$-dimensional approach should make it possible, in future, to understand theories whose $(1+1)$-dimensional counterparts are not already understood. On the other hand, a mathematically precise statement of the majority of results of this paper is given at the beginning of §4. This discussion is essentially self-contained.

In §1, we present a detailed, although elementary, exposition of the basic concepts of Kähler quantization of affine spaces and their symplectic quotients. We define the desired connection abstractly. As an example, in the last subsection we show explicitly how for quantization of the quotient of a vector space by a lattice, the connection is the operator appearing in the heat equation for classical theta functions.

The remainder of the paper is devoted to making the results of §1 explicit in such a way that they essentially carry over to the gauge theory case. In §2 we discuss this case in detail and construct the desired connection in a notation that is probably most familiar to physicists.

In §3 we present a more precise and geometric formulation of the results of §2 in notation suitable for arbitrary affine symplectic quotients. We derive a formula for the connection that may be written intrinsically on $\mathscr{M}$. This derivation is, of course, only formal for the gauge theory problem.

In §4, we state and prove most of the main results. Using an ansatz suggested by the results of §§2 and 3 and properties of the intrinsic geometry of $\mathscr{M}$, we find a well-defined connection. The properties of the intrinsic geometry of $\mathscr{M}$ which we need follow from the local version of the families index theorem and geometric invariant theory. Using these properties, and one further fact, we show that the connection is projectively flat. (Actually, there are several candidates for the "further fact" in question. One argument uses a global result—the absence of holomorphic vector fields on $\mathscr{M}$—while a second argument is based on a local differential geometric identity proved in §7. It should also be possible to make a third proof on lines sketched at the end of §6.) This section is the core section analyzing the properties of the connection and is rigorous since all the required analysis has already been done in the proof of the index theorem.

In §5 we shall concentrate on the gauge theory case when Σ is a torus. We give explicit formulas for our connection and a basis of parallel sections of the quantum Hilbert bundle $\mathscr{H}(\Sigma)$. We also show directly that our connection is unitary and has the curvature claimed. The parallel sections are identified with the Weyl-Kac characters for the representations of the loop group of G. This result is natural from the conformal field theory point of view, and was originally discussed from the point of view of quantization of Chern-Simons gauge theory in [7].

In §6 we make some preliminary comments about the unitarity of our connection.

In §7 we develop an extensive machinery allowing us to prove in a systematic way the one identity left unproved in §4. Our discussion, however, is incomplete in that we have not checked some details of the analysis of regularization.

The appendix contains further formulas relevant to §5.

We would like to thank M. Atiyah, V. Della Pietra, C. Fefferman, D. Freed, N. Hitchin, C. Simpson, and G. Washnitzer for helpful discussions.

1. Geometric setup and pushed down connection

In this section we consider the quantization of a finite-dimensional symplectic manifold $\mathscr{M}$ which is the symplectic quotient of an affine symplectic manifold $\mathscr{A}$ by a suitable subgroup of the affine symplectic group. Quantization of $\mathscr{M}$ is carried out by choosing a suitable complex structure J on $\mathscr{A}$ which induces one on $\mathscr{M}$. We describe the projectively flat connection whose existence shows that quantization of $\mathscr{M}$ is independent of the choice of J. This is an interesting, though fairly trivial, result about geometric quantization. Its real interest comes in the generalization to gauge theory, which will occupy the rest of the paper.

Most of this section is a review of concepts which are well known, although possibly not in precisely this packaging [22], [36], [37], [43]. We review this material in some detail in the hope of making the paper accessible.

1a. Symplectic geometry and Kähler quantization. To begin with, we consider a symplectic manifold $\mathscr{A}$, that is, a manifold with a closed and nondegenerate two-form ω. Nondegeneracy means that if we regard ω as a map from $\omega : T\mathscr{A} \to T^*\mathscr{A}$, then there is an inverse map ω^{-1} : $T^*\mathscr{A} \to T\mathscr{A}$. In local coordinates, a^i, if

$$\omega = \omega_{ij}\, da^i \wedge da^j, \tag{1.1}$$

and

$$\omega^{-1} = \omega^{ij} \frac{\partial}{\partial a^i} \otimes \frac{\partial}{\partial a^j}, \tag{1.2}$$

then the matrices ω_{ij} and ω^{ij} are inverses,

$$\omega_{ij}\omega^{jk} = \delta_i^{\ k}. \tag{1.3}$$

Let $C^\infty(\mathscr{A})$ denote the smooth functions on $\mathscr{A}$. Given $h \in C^\infty(\mathscr{A})$, we form the vector field $V_h = \omega^{-1}(dh)$ called the flow of h. It is a symplectic vector field—that is, the symplectic form ω is annihilated by the Lie derivative $\mathscr{L}_{V_h}$—since $\mathscr{L}_{V_h}(\omega) = (i_{V_h} d + d i_{V_h})\omega = d(i_{V_h}\omega)$ (since ω is closed) and by the definition of V_h one has $i_{V_h}(\omega) = -dh$. Conversely, given a symplectic vector field V, that is a vector field V such that $\mathscr{L}_V(\omega) = 0$, one has a closed one-form $\alpha = i_V(\omega)$. A function h such that $\alpha = dh$ is called a Hamiltonian function or moment map for V. If the first Betti number of $\mathscr{M}$ is zero, then every symplectic vector field on $\mathscr{A}$ can be derived from some Hamiltonian function.

The symplectic vector fields on $\mathscr{A}$ form a Lie algebra. If two symplectic vector fields V_f and V_g can be derived from Hamiltonian functions f and g, then their commutator $[V_f, V_g]$ can likewise be derived from a Hamiltonian function; in fact

$$[V_f, V_g] = V_{[f, g]_{\mathrm{PB}}}, \tag{1.4}$$

where $[f, g]_{\mathrm{PB}}$ denotes the so-called Poisson bracket

$$[f, g]_{\mathrm{PB}} = \omega^{-1}(df, dg) = \omega^{ij}\partial_i f \cdot \partial_j g. \tag{1.5}$$

Therefore, the symplectic vector fields that can be derived from Hamiltonians form a Lie subalgebra of the totality of symplectic vector fields.

Essentially by virtue of (1.4) the Poisson bracket obeys the Jacobi identity

$$[[f\,,\,g]_{\rm PB}\,,\,h]_{\rm PB}+[[g\,,\,h]_{\rm PB}\,,\,f]_{\rm PB}+[[h\,,\,f]_{\rm PB}\,,\,g]_{\rm PB}=0\,, \tag{1.6}$$

so that under the $[\ ,\ ]_{\rm PB}$ operation, the smooth functions on $\mathscr{A}$ have a Lie algebra structure. It is evident that the center of this Lie algebra consists of functions f such that $df = 0$; in other words, if $\mathscr{A}$ is connected, it consists of the constant functions. The smooth functions on $\mathscr{A}$ are also a commutative, associative algebra under ordinary pointwise multiplication, and the two structures are compatible in the sense that

$$[f\,,\,gh]_{\rm PB}=[f\,,\,g]_{\rm PB}\cdot h+[f\,,\,h]_{\rm PB}\cdot g. \tag{1.7}$$

These compatible structures $[\ ,\ ]_{\rm PB}$ and pointwise multiplication give $C^\infty(\mathscr{A})$ a structure of "Poisson-Lie algebra".

According to quantum mechanics textbooks, "quantization" of a symplectic manifold $\mathscr{A}$ means constructing "as nearly as possible" a unitary Hilbert space representation of the Poisson-Lie algebra $C^\infty(\mathscr{A})$. This would mean finding a Hilbert space H and a linear map $f\to\widehat{f}$ from smooth real-valued functions on $\mathscr{A}$ to selfadjoint operators on H such that $\widehat{(fg)}=\widehat{f}\cdot\widehat{g}$ and $\widehat{[f,g]}_{\rm PB}=i[\widehat{f}\,,\,\widehat{g}]$. One also requires (or proves from an assumption of faithfulness and irreducibility), that if 1 denotes the constant function on A, then $\widehat{1}$ is the identity operator on H.

This notion of what quantization should mean is however far too idealized; it is easy to see that such a Hilbert space representation of the Poisson-Lie algebra $C^\infty(\mathscr{A})$ does not exist. Quantum mechanics textbooks therefore instruct one to construct something that is "as close as possible" to a representation of $C^\infty(\mathscr{A})$. This is of course a vague statement. In general, a really satisfactory notion of what "quantization" should mean exists only in certain special classes of examples. The proper study of these examples, on the other hand, leads to much information. The examples we will be considering in this paper are affine spaces and their symplectic quotients by subgroups of the affine symplectic group obeying certain restrictions.

Prequantization. If one considers $C^\infty(\mathscr{A})$ purely as a Lie algebra, a Hilbert space representation can be constructed via the process of "prequantization".

Actually, for prequantization one requires that $\frac{1}{2\pi}\omega$ represents an integral cohomology class. This condition ensures the existence of a Hermitian line bundle $\mathscr{L}$ over $\mathscr{A}$ with a connection ∇ that is compatible with the Hermitian metric $\langle\ ,\ \rangle_{\mathscr{L}}$ and has curvature $-i\omega$. The isomorphism class of $\mathscr{L}$ (as a unitary line bundle with connection) may not be unique; given one choice of $\mathscr{L}$, any other choice is of the form $\mathscr{L}' = \mathscr{L} \otimes S$, where S is a flat unitary line bundle, determined by an element of $H^1(\mathscr{A}, U(1))$. The problem of prequantization has a solution for every choice of $\mathscr{L}$.

Let $\mathscr{C}$ be the group of diffeomorphisms of the total space of the line bundle $\mathscr{L}$ which preserves all the structure we have introduced (the fibration over $\mathscr{A}$, the connection, and the Hermitian structure). Let $H_{L^2}(\mathscr{A}, \mathscr{L})$ be the "prequantum" Hilbert space of all square integrable sections of $\mathscr{L}$. Since $\mathscr{C}$ acts on $\mathscr{L}$, it also acts on $H_{L^2}(\mathscr{A}, \mathscr{L})$. An element, D, of the Lie algebra of $\mathscr{C}$ is just a vector field on $\mathscr{M}$ lifted to act on $\mathscr{L}$. Acting on $H_{L^2}(\mathscr{A}, \mathscr{L})$, this corresponds to a first-order differential operator,

$$D = \nabla_T + ih. \tag{1.8}$$

Here T is the vector field representing the action of D on the base space $\mathscr{A}$ and h is a function on $\mathscr{A}$. The conditions that D preserves the connection is that for any vector field v we have

$$[\nabla_T + ih, \nabla_v] = \nabla_{\mathscr{L}_T v}. \tag{1.9}$$

Since the curvature of ∇ is $-i\omega$, this is true if and only if $T = V_h$. One may easily check that the map ρ_{pr} from $C^\infty(\mathscr{A})$ to the Lie algebra of $\mathscr{C}$ defined by

$$\rho_{\text{pr}}(h) = \tfrac{1}{i}\nabla_{V_h} + h \tag{1.10}$$

is an isomorphism of Lie algebras. In addition, the function 1 maps to the unit operator.

Prequantization, as just described, is a universal recipe which respects the Lie algebra structure of $C^\infty(\mathscr{A})$ at the cost of disregarding the other part of the Poisson-Lie structure, coming from the fact that $C^\infty(\mathscr{A})$ is a commutative associative algebra under multiplication of functions. Quantization, as opposed to prequantization, is a compromise between the two structures, and in contrast to prequantization, there is no universal recipe for what quantization should mean. We now turn to the case of affine spaces, the most important case in which there is a good recipe.

Quantization of affine spaces. Let $\mathscr{A}$ be a $2n$-dimensional affine space, with linear coordinates a^i, $i = 1 \ldots 2n$ and an affine symplectic structure

$$\omega = \omega_{ij}\, da^i\, da^j, \tag{1.11}$$

with ω_{ij} being an invertible (constant) skew matrix.

The Poisson brackets of the linear functions a^i are

$$[a^i, a^j]_{\mathrm{PB}} = \omega^{ij}. \tag{1.12}$$

In contrast to prequantization, in which one finds a Lie algebra representation of all of $C^\infty(\mathscr{A})$ in a Hilbert space $\mathscr{H}$, in quantization we content ourselves with finding a Hilbert space representation of the Poisson brackets of the linear functions, that is, a Hilbert space representation of the Lie algebra

$$[\widehat{a^i}, \widehat{a^j}] = -i\omega^{ij}. \tag{1.13}$$

Actually, we want Hilbert space representations of the "Heisenberg Lie algebra" (1.13) that integrate to representations of the corresponding group. This group, the Heisenberg group, is simply the subgroup of $\mathscr{C}$ that lifts the affine translations. According to a classic theorem by Stone and von Neumann, the irreducible unitary representation of the Heisenberg group is unique up to isomorphism, the isomorphism being unique up to multiplication by an element of $U(1)$ (the group of complex numbers of modulus one). Representing only the Heisenberg group—and not all of $C^\infty(\mathscr{A})$—means that quantization can be carried out in a small subspace of the prequantum Hilbert space.

Action of the affine symplectic group. Before actually constructing a representation of the Heisenberg group, let us discuss some properties that any such representation must have.

The affine symplectic group $\mathscr{W}$—the group of affine transformations of $\mathscr{A}$ that preserve the symplectic structure—acts by outer automorphisms on the Lie algebra (1.13). The pullback of a representation ρ of (1.13) by an element w of $\mathscr{W}$ is another representation ρ' of (1.13) in the same Hilbert space H. The uniqueness theorem therefore gives a unitary operator $U(w) : H \to H$ such that $\rho' = U(w) \circ \rho$. The $U(w)$ are unique up to multiplication by an element of $U(1)$, and therefore it is automatically true that for $w, w' \in H$, $U(ww') = U(w)U(w')\alpha(w, w')$ where $\alpha(w, w')$ is a $U(1)$-valued two-cocycle of $\mathscr{W}$. Thus, the $U(w)$ give a representation of a central extension by $U(1)$ of the group $\mathscr{W}$. It can be shown that if one restricts to the *linear* symplectic group—the subgroup of $\mathscr{W}$ that fixes a point in $\mathscr{A}$—then (for finite-dimensional affine spaces) the kernel of this central extension can be reduced to $\mathbb{Z}/2\mathbb{Z}$.

Now, let us investigate the extent to which a representation ρ of the Lie algebra (1.13) can be extended to a representation of the Poisson-Lie algebra $C^\infty(\mathscr{A})$. One immediately sees that this is impossible, since one would require both $\rho(a^i a^j) = \rho(a^i)\rho(a^j)$ and $\rho(a^i a^j) = \rho(a^j a^i) = \rho(a^j)\rho(a^i)$; but

$$\rho(a^i)\rho(a^j) - \rho(a^j)\rho(a^i) = -i\omega^{ij}. \tag{1.14}$$

Thus, ρ cannot be extended to a representation of $C^\infty(\mathscr{A})$. However, the right-hand side of (1.14), though not zero, is in the center of $C^\infty(\mathscr{A})$, and this enables us to take one more important step. Defining

$$\rho(a^i a^j) = \tfrac{1}{2}(\rho(a^i)\rho(a^j) + \rho(a^j)\rho(a^i)), \tag{1.15}$$

one verifies that

$$\begin{aligned} [\rho(a^i a^j), \rho(a^k a^l)] &= -i\rho([a^i a^j, a^k a^l]_{\text{PB}}) \\ [\rho(a^i), \rho(a^k a^l)] &= -i\rho([a^i, a^k a^l]_{\text{PB}}). \end{aligned} \tag{1.16}$$

To interpret (1.16) observe that the linear and quadratic functions on $\mathscr{A}$ form, under Poisson bracket, a Lie algebra which is a central extension of the Lie algebra of the affine symplectic group. In other words, the Hamiltonian functions from which the generators of the affine symplectic group can be derived are simply the linear and quadratic functions on $\mathscr{A}$. Equation (1.16), together with (1.13), means that *any* representation of the Lie algebra (1.13) automatically extends to a projective representation of the Lie algebra of the affine symplectic group $\mathscr{W}$. This is an infinitesimal counterpart of a fact that we have already noted: by virtue of the uniqueness theorems for irreducible representations of (1.13), the group $\mathscr{W}$ automatically acts projectively in any such representation.

The verification of (1.16) depends on the fact that the ambiguity in the definition of $\rho(a^i a^j)$—the difference between $\rho(a^i)\rho(a^j)$ and $\rho(a^j)\rho(a^i)$—is central. For polynomials in the a^i of higher than second order, different orderings differ by terms that are no longer central, and it is impossible to extend ρ to a representation of $C^\infty(\mathscr{A})$ even as a Lie algebra, let alone a Poisson-Lie algebra. It is natural to adopt a symmetric definition

$$\rho(a^{i_1}a^{i_2}\cdots a^{i_n}) = \frac{1}{n!}(a^{i_1}a^{i_2}\cdot a^{i_n} + \text{permutations}), \tag{1.17}$$

but for $n > 2$ this does not give a homomorphism of Lie algebras.

Quantization. There remains now the problem of actually constructing Hilbert space representations of (1.13). There are two standard constructions (which are equivalent, of course, in view of the uniqueness theorem). Each construction involves a choice of a "polarization", that is, a maximal linearly independent commuting subset of the linear functions on $\mathscr{A}$. In the first approach, one takes these functions to be real valued. In the second approach, they are complex valued and linearly independent over $\mathbb{C}$.

We will describe the second approach; it is the approach that will actually be useful in what follows.

Pick a complex structure J on $\mathscr{A}$, invariant under affine translations, such that ω is positive and of type $(1, 1)$. Then one can find n linear functions z^i that are holomorphic in the complex structure J such that

$$\omega = +i dz^i \wedge d\overline{z}^i. \tag{1.18}$$

Let $\mathscr{L}$ be the prequantum line bundle introduced in our discussion of prequantization. We recall that $\mathscr{L}$ is to be a Hermitian line bundle with a connection ∇ whose curvature form is $-i\omega$. Since the $(0, 2)$ part of ω vanishes, the connection ∇ gives $\mathscr{L}$ a structure as a holomorphic line bundle. In fact, $\mathscr{L}$ may be identified as the trivial holomorphic line bundle whose holomorphic sections are holomorphic functions ψ and with the Hermitian structure $|\psi|^2 = \exp(-h) \cdot \overline{\psi}\psi$, with $h = \sum_i \overline{z}^i z^i$. Indeed, the connection ∇ compatible with the holomorphic structure and with this Hermitian structure has curvature $\overline{\partial}\partial(-h) = \sum dz^i d\overline{z}^i = -i\omega$. Since $H^1(\mathscr{A}, U(1)) = 0$, the prequantum line bundle just constructed is unique up to isomorphism.

We now define the quantum Hilbert space $\mathscr{H}_Q|_J$, in which the Heisenberg group is to be represented, to be the Hilbert space $H^0_{L^2}(\mathscr{A}, \mathscr{L})$ of *holomorphic* L^2 sections of $\mathscr{L}$. We recall that, by contrast, the prequantum Hilbert space consists of all L^2 sections of $\mathscr{L}$ without the holomorphicity requirement.

The required representation ρ of the Heisenberg group is the restriction of the prequantum action to the quantum Hilbert space. At the Lie algebra level, the z^i act as multiplication operators,

$$\rho(z^i)\psi = z^i\psi, \tag{1.19}$$

and the $\overline{z}^i$ act as derivatives with respect to the z^i,

$$\rho(\overline{z}^i)\psi = \frac{\partial}{\partial z^i}\psi. \tag{1.20}$$

That this representation is unitary follows from the identity

$$\langle z^i\chi, \psi\rangle = \left\langle \chi, \frac{\partial}{\partial z^i}\psi \right\rangle, \tag{1.21}$$

which asserts that $\rho(\overline{z}^i)$ is the Hermitian adjoint of $\rho(z^i)$. (Of course, with the chosen Hermitian structure on $\mathscr{L}$, $\langle \chi, \psi\rangle = \int \exp(-\sum_i \overline{z}_i z_i) \cdot \overline{\chi}\psi$.)

Irreducibility of this representation of the Heisenberg group can be proved in an elementary fashion. This irreducibility is a hallmark of quantization as opposed to prequantization.

Because of the uniqueness theorem for irreducible unitary representations of the Heisenberg group, the Hilbert space $\mathscr{H}_Q|_J$ that we constructed above is, up to the usual projective ambiguity, independent of the choice of J (as long as J obeys the restrictions we imposed: it is invariant under the affine translations, and ω is positive and of type $(1, 1)$). It automatically admits a projective action of the group of all affine symplectic transformations, including those that do not preserve J.

Infinitesimally, the independence of J is equivalent to the existence of a projective action of the Lie algebra of the affine symplectic group. Its existence follows from what we have said before; we have noted in (1.16) that given *any* representation $x^i \to \rho(x^i)$ of the Heisenberg Lie algebra, no matter how constructed, one can represent the Lie algebra of the affine symplectic group by expressions quadratic in the $\rho(x^i)$. In the representation that we have constructed of the Heisenberg Lie algebra, since the $\rho(z^i)$ and $\rho(\overline{z}^i)$ are differential operators on $\mathscr{A}$ of order 0 and 1, respectively, and the Lie algebra of the affine symplectic group is represented by expressions quadratic in these, this Lie algebra is represented by differential operators of (at most) second order. By the action of this Lie algebra, one sees the underlying symplectic geometry of the affine space $\mathscr{A}$, though an arbitrary choice of a complex structure J of the allowed type has been used in the quantization.

Quantization of Kähler manifolds. In this form, one can propose to "quantize" symplectic manifolds more general than affine spaces. Let $(\mathscr{A}, \omega)$ be a symplectic manifold with a chosen complex structure J such that ω is positive and of type $(1, 1)$ (and so defines a Kähler structure on the complex manifold $\mathscr{M}$). Any prequantum line bundle $\mathscr{L}$ automatically has a holomorphic structure, since its curvature is of type $(1, 1)$, and the Hilbert space $H^0_{L^2}(\mathscr{A}, \mathscr{L})$ can be regarded as a quantization of $(\mathscr{A}, \omega)$. In this generality, however, Kähler quantization depends on the choice of J and does not exhibit the underlying symplectic geometry. What is special about the Kähler quantization of affine spaces is that in that case, through the action of the affine symplectic group, one can see the underlying symplectic geometry even though a complex structure is used in quantization.

Most of this paper will in fact be concerned with quantization of special Kähler manifolds that are closely related to affine spaces. So we will now discuss Kähler quantization in detail, considering first some general features and then properties that are special to affine spaces. To begin, we review some basic definitions to make our notation clear.

An almost complex structure J on a manifold $\mathscr{A}$ is a linear operator from $T\mathscr{A}$ to itself with $J^2 = -1$, i.e. a complex structure on $T\mathscr{A}$. On $T\mathscr{A} \otimes \mathbb{C}$ we can form the projection operators $\pi_z \equiv \frac{1}{2}(1 - iJ)$ and $\pi_{\bar{z}} \equiv \frac{1}{2}(1 + iJ)$. The image of π_z is the subspace of $T\mathscr{A} \otimes \mathbb{C}$ on which J acts by multiplication by i. It is called $T^{(1,0)}\mathscr{A}$ or the holomorphic tangent space. Similarly, $T^{(0,1)}\mathscr{A}$ is the space on which J acts by multiplication by $-i$. The transpose maps π_z^{T} and $\pi_{\bar{z}}^{\mathrm{T}}$ act on $T^*\mathscr{A} \otimes \mathbb{C}$. We define $T^{*(1,0)}$ and $T^{*(0,1)}$ as their images. Given local coordinates a^i, we may define

$$da^{\underline{i}} = \pi_z(da^i), \qquad da^{\bar{i}} = \pi_{\bar{z}} da^i. \tag{1.22}$$

The statement that J is a complex structure means that we may pick our coordinates a^i so that $da^{\underline{i}}$ actually is the differential of a complex valued function $a^{\underline{i}}$ and $da^{\bar{i}}$ is the differential of the complex conjugate $a^{\bar{i}}$.

We should make contact with more usual notation. The usual complex and real coordinates are

$$\begin{gathered} z^i = x^i + iy^i \quad \text{for } i = 1, \dots, n, \\ a^i = \begin{cases} x^i & \text{for } i = 1, \dots, n, \\ y^{i-n} & \text{for } i = n+1, \dots, 2n. \end{cases} \end{gathered} \tag{1.23}$$

So we have

$$a^{\underline{i}} = \begin{cases} \frac{1}{2} z^i & \text{for } i = 1, \dots, n, \\ -\frac{i}{2} z^{i-n} & \text{for } i = n+1, \dots, 2n. \end{cases} \tag{1.24}$$

We may decompose a 2-form σ as the sum of its $(2, 0)$, $(1, 1)$, and $(0, 2)$ components:

$$\begin{gathered} \sigma^{(2,0)} = \sigma_{\underline{i}\underline{j}} da^{\underline{i}} da^{\underline{j}}, \qquad \sigma^{(0,2)} = \sigma_{\bar{i}\bar{j}} da^{\bar{i}} da^{\bar{j}}, \\ \sigma^{(1,1)} = \sigma_{\underline{i}\bar{j}} da^{\underline{i}} da^{\bar{j}} + \sigma_{\bar{i}\underline{j}} da^{\bar{i}} da^{\underline{j}}. \end{gathered} \tag{1.25}$$

In general, any real tensor can be thought of as a complex tensor with the indices running over $\underline{i}$ and $\bar{i}$ which correspond to a basis for $T\mathscr{A} \otimes \mathbb{C}$.

We also assume that J is compatible with ω in the sense that $\omega(Jv, Jw) = \omega(v, w)$ for any $v, w \in T\mathscr{A}$. This amounts to the assumption that

$$\begin{gathered} J^T \omega = -\omega J \\ J^j{}_i \omega_{jk} = -\omega_{ij} J^j{}_k. \end{gathered} \tag{1.26}$$

This is so exactly when ω is purely of type $(1, 1)$.

We may form the map $g = \omega \circ J$ from $T\mathscr{A}$ to $T^*\mathscr{A}$. Equivalently g is the J compatible nondegenerate symmetric bilinear form:

$$g(v, w) = \omega(v, Jw) \quad \text{for } v, w \in \mathscr{A}. \tag{1.27}$$

Finally, we assume that J is chosen so that g is a positive definite metric. In summary, $T\mathscr{A}$ is a complex manifold with a Riemannian metric, g, which is compatible with the complex structure and so that $\omega = -g \circ J$ is a symplectic form. This is just the definition of a Kähler manifold; ω is also called the Kähler form.

A connection ∇ on a vector bundle $\mathscr{V}$ over a Kähler manifold which obeys the integrability condition

$$0 = [\nabla_{\bar{i}}, \nabla_{\bar{j}}] \tag{1.28}$$

induces a holomorphic structure on $\mathscr{V}$, the local holomorphic sections being the sections annihilated by $\nabla_{\bar{i}}$. In particular, since ω is of type $(1, 1)$, the prequantum line bundle $\mathscr{L}$, which is endowed with a unitary connection obeying

$$[\nabla_i, \nabla_j] = -i\omega_{ij}, \tag{1.29}$$

is always endowed with a holomorphic structure. It is this property that enables one to define the quantum Hilbert space $\mathscr{H}_Q|_J$ as $H^0_{L^2}(\mathscr{A}, \mathscr{L})$.

Variation of complex structure. In general, given a symplectic manifold $\mathscr{A}$ with symplectic structure ω, it may be impossible to find a Kähler polarization—that is, a complex structure J for which ω has the properties of a Kähler form. If however a Kähler polarization exists, it is certainly not unique, since it can be conjugated by any symplectic diffeomorphism. To properly justify the name "quantization", which implies a process in which one is seeing the underlying symplectic geometry and not properties that depend on the choice of a Kähler structure, one would ideally like to have a canonical identification of the $\mathscr{H}_Q|_J$ as J varies. This, however, is certainly too much to hope for.

In many important problems, there is a natural choice of Kähler polarization—for instance, a unique choice compatible with the symmetries of the problem. We will be dealing with situations in which there is not a single natural choice of Kähler polarization, but a preferred family $\mathscr{T}$. For instance, for $\mathscr{A}$ affine we take $\mathscr{T}$ to consist of translationally invariant complex structures such that ω is a Kähler form. In such a case, the spaces $\mathscr{H}_Q|_J = H^0_{L^2}(\mathscr{A}, \mathscr{L})$ are the fibers of a Hilbert bundle $\mathscr{H}_Q$ over $\mathscr{T}$. $\mathscr{H}_Q$ is a subbundle of the trivial Hilbert bundle with total space

$\mathcal{H}_{\text{pr}} = H_{L^2}(\mathcal{A}, \mathcal{L}) \times \mathcal{T}$. We will aim to find a canonical (projective) identification of the fibers $\mathcal{H}_Q|_J$, as J varies, by finding a natural projectively flat Hermitian connection $\delta^{\mathcal{H}_Q}$ on $\mathcal{H}_Q$. The parameter spaces $\mathcal{T}$ will be simply connected, so such a connection leads by parallel transport to an identification of the fibers of $\mathcal{H}_Q$.

Let $\mathcal{C}'$ be the subgroup of $\mathcal{C}$ consisting of those elements whose action on the space of complex structures on $\mathcal{A}$ leaves $\mathcal{T}$ invariant setwise. An element $\phi \in \mathcal{C}'$ maps $\mathcal{H}_Q|_J$ to $\mathcal{H}_Q|_{\phi J}$ in an obvious way. Using the projectively flat connection $\delta^{\mathcal{H}^Q}$ to identify $\mathcal{H}_Q|_{\phi J}$ with $\mathcal{H}_Q|_J$, we get a unitary operator $\phi|_J : \mathcal{H}_Q|_J \to \mathcal{H}_Q|_J$. We consider the association $\phi \to \phi|_J$ to represent a quantization of the symplectic transformation ϕ if the $\phi|_J$ are invariant (at least projectively) under parallel transport by $\delta^{\mathcal{H}^Q}$. In this case we say that ϕ is quantizable. It is evident that the symplectic transformations that are quantizable in this sense form a group $\mathcal{C}''$; for any J, $\phi \to \phi|_J$ is a projective representation of this group (the representations obtained for different J's are of course conjugate under parallel transport by $\delta^{\mathcal{H}^Q}$).

Repeating this discussion at the Lie algebra level, we obtain the following definition of quantization of functions h whose flow leaves $\mathcal{T}$ invariant. (For $\mathcal{A}$ affine, h is any quadratic function.) Let $\underset{h}{\delta} J = \mathcal{L}_{V_h}(J) \in T_J\mathcal{T}$ be the infinitesimal change in J induced by h. Let δ be the trivial connection on the trivial bundle $\mathcal{H}_{\text{pr}}$. The quantization of h can be written as a sum of first-order differential operators on $\Gamma(\mathcal{T}, \mathcal{H}_{\text{pr}})$,

$$i\hat{h} = i\rho_{\text{pr}}(h) + \delta_{\underset{h}{\delta} J} - \delta^{\mathcal{H}_Q}_{\underset{h}{\delta} J} + \text{constant}. \tag{1.30}$$

The first term is the naive prequantum contribution. The second term represents the fact that the prequantum operator should also be thought of as moving the complex structure. The third term is our use of $\delta^{\mathcal{H}_Q}$ to return to the original complex structure so that $\hat{h}$ is just a linear transformation on the fibers of $\mathcal{H}_{\text{pr}}$. To check that (1.30) leaves the subbundle $\mathcal{H}_Q$ invariant we observe that acting on sections of $\mathcal{H}_Q$

$$\begin{aligned} \nabla_{\pi_{\bar{z}}v} \circ \hat{h} &= [\nabla_{\pi_{\bar{z}}v}, \rho_{\text{pr}}(h) - i\delta_{\underset{h}{\delta} J}] \\ &= -\nabla_{-i\mathcal{L}_{V_h}(\pi_{\bar{z}}v) + i(\delta_{\underset{h}{\delta} J})(\pi_{\bar{z}}v)} = 0. \end{aligned} \tag{1.31}$$

In the first line of (1.31), $\delta_{\underset{h}{\delta} J}$ is the trivial connection acting in the $\underset{h}{\delta} J$ direction on sections of the trivial bundle $\mathcal{H}_{\text{pr}} \to \mathcal{T}$. In the second line it is the trivial connection on $T\mathscr{A} \otimes \mathbb{C} \times \mathscr{T} \to \mathscr{T}$. The first equality in (1.31) follows from the facts that $\pi_{\bar{z}}v$ annihilates holomorphic sections and that $\delta^{\mathscr{H}_Q}$ takes holomorphic sections to holomorphic sections. The second equality follows from (1.9). Equation (1.31) shows that $\hat{h}$ preserves holomorphicity as desired.

The requirement that $\hat{h}$ is independent of complex structure is the statement that

$$\delta^{\mathscr{H}_Q}\hat{h} = 0. \tag{1.32}$$

If $\delta^{\mathscr{H}_Q}$ is projectively flat this implies that quantization is a projective representation:

$$\widehat{[h_1, h_2]_{\mathrm{PB}}} = i[\hat{h}_1, \hat{h}_2] + \text{constant}. \tag{1.33}$$

Connection for quantization of affine space. We now turn to the case in which $\mathscr{A}$ is an affine symplectic manifold and $\mathscr{T}$ consists of translationally invariant complex structures. We take a^i to be global affine coordinates. By the uniqueness theorem for irreducible unitary representations of the Heisenberg algebra we know that the projectively flat connection $\delta^{\mathscr{H}_Q}$ must exist. It may be defined in several equivalent ways which we shall discuss in turn.

1. We first present a simple explicit formula for $\delta^{\mathscr{H}_Q}$ and then show that it corresponds to any of the definitions below. The connection $\delta^{\mathscr{H}_Q}$ is given by

$$\delta^{\mathscr{H}_Q} = \delta - \mathscr{O}^{up}, \tag{1.34.1}$$

$$\mathscr{O}^{up} = M^{\underline{i}\underline{j}}\nabla_{\underline{i}}\nabla_{\underline{j}} \quad \text{with} \quad M^{\underline{i}\underline{j}} = -\tfrac{1}{4}(\delta J\omega^{-1})^{\underline{i}\underline{j}}. \tag{1.34.2}$$

Here δJ is a one form on $\mathscr{T}$ with values in $\mathrm{Hom}(T\mathscr{A}, T\mathscr{A})$. We call $\mathscr{O}^{up}$ the connection one-form for $\delta^{\mathscr{H}_Q}$. It is a second-order differential operator on $\mathscr{A}$ acting on sections of $\mathscr{L}$. We use the superscript '*up*' to distinguish it from the connection one-form which we will construct for quantization of the symplectic quotient $\mathscr{M}$.

To demonstrate that $\delta^{\mathscr{H}_Q}$ preserves holomorphicity and is projectively flat, we need the variation with respect to J of the statements that $J^2 = -1$ and that J is ω-compatible (1.26), that is,

$$0 = J\delta J + \delta J J = J^i{}_j\delta J^j{}_k + \delta J^i{}_j J^j{}_k = 2i\delta J^{\underline{i}}{}_{\underline{k}} - 2i\delta J^{\bar{i}}{}_{\bar{k}} \tag{1.35}$$

$$(\omega\delta J)_{ij} = (\omega\delta J)_{\underline{i}\underline{j}} + (\omega\delta J)_{\bar{i}\bar{j}} \quad \text{is symmetric.} \tag{1.36}$$

Using these identities as well as the fact that ∇ has curvature $-i\omega$, it is easy to check that $\delta^{\mathscr{H}_Q}$ preserves holomorphicity (so that it does in fact give a connection on the bundle $\mathscr{H}_Q$ over $\mathscr{T}$). To calculate the curvature $R^{\delta^{\mathscr{H}_Q}} = (\delta^{\mathscr{H}_Q})^2$ we first observe that $\mathscr{O}^{up} \wedge \mathscr{O}^{up} = 0$ because holomorphic derivatives commute. To calculate $\delta\mathscr{O}^{up}$ we must remember that the meaning of the indices $\underline{i}$ and $\bar{i}$ change as we change J. One way to account for this is to use only indices of type i and explicitly write π_z wherever needed. Using the formulas

$$\delta\pi_z = -\frac{i}{2}\delta J \tag{1.37.1}$$

$$\delta\pi_{\bar{z}} = +\frac{i}{2}\delta J, \tag{1.37.2}$$

we find

$$R^{\delta^{\mathscr{H}_Q}} = -\delta\mathscr{O}^{up} = -\tfrac{1}{8}\delta J^{\underline{i}}{}_{\bar{j}}\delta J^{\bar{j}}{}_{\underline{i}} = -\tfrac{1}{8}\mathrm{Tr}(\pi_z\delta J \wedge \delta J). \tag{1.38}$$

This is a two-form on $\mathscr{T}$ whose coefficients are multiplication operators by constant functions, i.e., $\delta^{\mathscr{H}_Q}$ is projectively flat as desired.

2. The essential feature of the connection that we have just defined is that $\mathscr{O}^{up} = \delta - \delta^{\mathscr{H}^Q}$ is a second-order differential operator. The reason for this key property is that in quantization of an affine space, the Lie algebra of the affine symplectic group is represented by second-order differential operators. Indeed, a change δJ of complex structure is induced by the flow of the Hamiltonian function

$$h = -\tfrac{1}{4}(\omega J\delta J)_{ij}a^i a^j. \tag{1.39}$$

In the discussion leading to (1.15), we have already defined the quantization of a quadratic function

$$h = h_{ij}a^i a^j + h_i a^i + h_0 \tag{1.40}$$

by symmetric ordering,

$$\rho(h) = \hat{h} = h_{ij}\tfrac{1}{2}\{\hat{a}^i\hat{a}^j + \hat{a}^j\hat{a}^i\} + h_i\hat{a}^i + h_0. \tag{1.41}$$

This preserves holomorphicity for any complex structure J and gives a representation of the quadratic Hamiltonian functions on $\mathscr{H}_Q|_J$. According to (1.30) (dropping the constant), $\mathscr{O}^{up} = \delta - \delta^{\mathscr{H}^Q}$ is to be simply $i\hat{h} - i\rho_{\mathrm{pr}}(h)$, with h in (1.41). This leads to the definition (1.34) of the connection $\delta^{\mathscr{H}_Q}$.[2]

[2]Note that (1.30), with a constant included, holds true for arbitrary h, and not just those of the form (1.39). By properly including the "metaplectic correction" we can actually find a flat connection and arrange for all unwanted constant factors to vanish.

3. One natural candidate, which exists quite generally, for the connection on $\mathcal{H}_Q$ is the "orthogonally projected" connection for $\mathcal{H}_Q$ considered as a subbundle of the trivial Hilbert bundle $\mathcal{H}_{\mathrm{pr}}$. It may be defined by

$$< \psi|\delta^{\mathcal{H}_Q}\psi' > = < \psi|\delta\psi' > \quad \text{for } \psi, \psi' \in \Gamma(\mathcal{T}, \mathcal{H}_Q). \tag{1.42}$$

Although we may write this formula down quite generally, there is no general reason that it should yield a projectively flat connection. However, in the case at hand, we may check that this definition agrees with that of points 1 and 2 above. This is so because $\mathcal{O}^{up}$ does not change form after integrating by parts, and $\mathcal{O}^{up}\overline{\psi} = 0$. This implies that our connection is in fact unitary.

4. Closely related to point 3 is the fact that $\delta^{\mathcal{H}_Q}$ may also be described as the unique unitary connection compatible with the holomorphic structure on $\mathcal{H}_Q$. We discussed the holomorphic structure on $\mathcal{H}_Q$ previously. To describe it explicitly we first note that a complex structure on the space $\mathcal{T}$ is defined by stating that the forms $\delta J^{j}{}_{\bar{i}}$ are of type $(1, 0)$ and that the forms $\delta J^{\bar{j}}{}_{\underline{i}}$ are of type $(0, 1)$. In other words, the $(1, 0)$ and $(0, 1)$ pieces of δJ are

$$\delta J^{(1,0)} = \pi_z \delta J \pi_{\bar{z}} \quad \text{and} \quad \delta J^{(0,1)} = \pi_{\bar{z}} \delta J \pi_z. \tag{1.43}$$

Let $\delta^{(1,0)}$ and $\delta^{(0,1)}$ be the holomorphic and antiholomorphic pieces of the trivial connection on $\mathcal{H}_{\mathrm{pr}}$. Holomorphic sections of $\mathcal{H}_{\mathrm{pr}}$ are those sections which are annihilated by the $\overline{\partial}$ operator $\delta^{(0,1)}$. We define sections of $\mathcal{H}_Q$ to be holomorphic if they are holomorphic as sections of $\mathcal{H}_{\mathrm{pr}}$. The integrability condition that we can find a local holomorphic trivialization of $\mathcal{H}_Q$ is satisfied if we can show that $\delta^{(0,1)}$ leaves $\mathcal{H}_Q$ invariant. But this is true since for ψ a section of $\mathcal{H}_Q$, we have

$$\nabla_{\bar{k}} \delta^{(0,1)} \psi = [\pi^{j}_{\bar{z}k} \nabla_j, \delta^{(0,1)}]\psi = -\tfrac{i}{2}(\delta J^{(0,1)})^{\bar{j}}{}_{\underline{k}} \nabla_{\bar{j}} \psi = 0. \tag{1.44}$$

The statement that $\delta^{\mathcal{H}_Q}$ as defined in point 1 above is compatible with the holomorphic structure is just the observation that $\delta^{\mathcal{H}_Q(0,1)} = \delta^{(0,1)}$ since $\mathcal{O}$ only depends on $\delta J^{(1,0)}$.

1b. Symplectic quotients and pushing down geometric objects. Affine spaces by themselves are comparatively dull. The facts just described get considerably more interest because they have counterparts for symplectic quotients of affine spaces. Our applications will ultimately come by considering finite-dimensional symplectic quotients of infinite-dimensional affine spaces.

Symplectic quotients. To begin, we discuss symplectic quotients of a general symplectic manifold $\mathscr{A}$, ω. Suppose that a group $\mathscr{G}$ acts on $\mathscr{A}$ by symplectic diffeomorphisms. We would like to define the natural notion of the "quotient" of a symplectic manifold by a symplectic group action. This requires defining the "moment map".

Let $\mathbf{g}$ be the Lie algebra of $\mathscr{G}$ and $T : \mathbf{g} \to \mathrm{Vect}(\mathscr{A})$ be the infinitesimal group action. Since $\mathscr{G}$ preserves ω, the image of T consists of symplectic vector fields. A comoment map for the $\mathscr{G}$ action is a $\mathscr{G}$ invariant map, F, from $\mathbf{g}$ to the Hamiltonian functions on $\mathscr{A}$ (where $\mathscr{G}$ acts on $\mathbf{g}$ by the adjoint action). To express this in component notation, let L_a be a basis for $\mathbf{g}$ and $T_a = T(L_a)$. Since T is a representation, we have

$$[T_a , T_b] = f_{ab}{}^c T_c , \tag{1.45}$$

where $f_{ab}{}^c$ are the structure constants of $\mathscr{G}$. The comoment map is given by functions F_a whose flow is T_a. Invariance of F under the connected component of $\mathscr{G}$ is equivalent to the statement that F is a Lie algebra homomorphism,

$$\{F_a , F_b\}_{\mathrm{PB}} = f_{ab}{}^c F_c. \tag{1.46}$$

For each $A \in \mathscr{A}$, $F_a(A)$ are the components of a vector in the dual space $\mathbf{g}^\vee$. We may view F as a map from $\mathscr{A}$ to $\mathbf{g}^\vee$. Viewed this way it is called a moment map.

Since the moment map and $\{0\} \subset \mathscr{G}$ are $\mathscr{G}$ invariant, so is $F^{-1}(0)$. The quotient space $\mathscr{M} = F^{-1}(0)/\mathscr{G}$ is called the symplectic or Marsden-Weinstein quotient of $\mathscr{A}$ by $\mathscr{G}$.[3] With mild assumptions, $\mathscr{M}$ is a non-singular manifold near the points corresponding to generic orbits of $\mathscr{G}$ in $F^{-1}(0)$. We will always restrict ourselves to nonsingular regions of $\mathscr{M}$, although we do not introduce any special notation to indicate this. We have the quotient map:

$$\begin{aligned} \pi : F^{-1}(0) &\to F^{-1}(0)/\mathscr{G} = \mathscr{M} \\ A &\mapsto \tilde{A}. \end{aligned} \tag{1.47}$$

We may define a symplectic structure, $\tilde{\omega}$, on $\mathscr{M}$ by

$$\tilde{\omega}_{\tilde{A}}(\tilde{v} , \tilde{w}) = \omega_A(v , w) \quad \text{for } \tilde{v} , \tilde{w} \in T_{\tilde{A}}\mathscr{M}. \tag{1.48}$$

[3]This quotient plays a role in elementary physics. If $\mathscr{A}$ is the phase space of a physical system, and $\mathscr{G}$ as a group of symmetries, then $\mathscr{M}$ is simply the phase space for the effective dynamics after one restricts to the level sets of the conserved momenta and solves the equations that can be integrated trivially due to group invariance. Alternatively, if the F_a are constraints generating gauge transformations of an unphysical phase space, then $\mathscr{M}$ is the physical phase space left after solving the constraints and identifying gauge equivalent configurations.

By $\mathcal{G}$-invariance and the fact that

$$\omega(T_a, u) = 0 \quad \text{for } u \in TF^{-1}(0)\,, \tag{1.49}$$

the definition is independent of our choice of A, v, and w. This is the first example of our theme of "pushing down" geometric objects from $\mathcal{A}$ to $\mathcal{M}$. The basic principle is that the objects (symplectic form, complex structure, bundles, connections, etc.) that we consider on $\mathcal{A}$ are $\mathcal{G}$-invariant so that when restricted to $F^{-1}(0)$ they push down to the corresponding objects on $\mathcal{M}$.

Pushing down the prequantum line bundle. In order to push down the prequantum line bundle we must assume we are given a lift of the $\mathcal{G}$-action on $\mathcal{A}$ to a $\mathcal{G}$-action on $\mathcal{L}$ which preserves the connection and Hermitian structure, i.e., an action by elements of $\mathcal{C}$. The Lie algebra version of such a lift is just a moment map. We may define the pushdown bundle $\tilde{\mathcal{L}}$ by stating its sections,

$$\Gamma(\mathcal{M}, \tilde{\mathcal{L}}) = \Gamma(F^{-1}(0), \mathcal{L})^{\mathcal{G}}. \tag{1.50}$$

The $\mathcal{G}$ superscript tells us to take the $\mathcal{G}$-invariant subspace. A line bundle on $\mathcal{M}$ with (1.50) as its sheaf of sections will exist if $\mathcal{G}$ acts freely on $F^{-1}(0)$ (or more generally if for all $x \in F^{-1}(0)$, the isotropy subgroup of x in $\mathcal{G}$ acts trivially on the fiber of $\mathcal{L}$ at x). A section $\psi \in \Gamma(F^{-1}(0), \mathcal{L})$ is invariant under the connected component of $\mathcal{G}$ precisely if

$$0 = i\rho(F_a)_{\mathrm{pr}} = \nabla_{T_a}\psi. \tag{1.51}$$

The pushdown connection may be defined by

$$\tilde{\nabla}_{\tilde{v}}\psi = \nabla_v\psi. \tag{1.52}$$

Here v is any vector field on $F^{-1}(0)$ which pushes forward to $\tilde{v}$ on $\mathcal{M}$. By (1.51) the right-hand side of (1.52) is independent of our choice of v. To show that (1.52) is a good definition we must show that the right-hand side is annihilated by ∇_{T_a}. This can be done using (1.29) and (1.49). Similarly, one can check that $\tilde{\nabla}$ has curvature $-i\tilde{\omega}$.

Pushing down the complex structure. To proceed further, we assume that $\mathcal{A}$ is an affine space and that $\mathcal{G}$ is a Lie subgroup of the affine symplectic group such that (i) there is an invariant metric on the Lie algebra $\mathbf{g}$, and (ii) the action of $\mathcal{G}$ on $\mathcal{A}$ leaves fixed an affine Kähler polarization. We continue to assume that the $\mathcal{G}$-action on $\mathcal{A}$ has been lifted to an action on $\mathcal{L}$ with a choice of moment map. We let $\mathcal{T}$ be the space of Kähler polarizations of $\mathscr{A}$ that are invariant under the affine translations and also $\mathscr{G}$-invariant. $\mathscr{T}$ is nonempty and contractible.

Since $\mathscr{G}$ acts linearly, there is a unique extension of the $\mathscr{G}$-action to an action of the complexification $\mathscr{G}_c$ which is holomorphic as a function from $\mathscr{G}_c \times \mathscr{A}$ to $\mathscr{A}$. In a closely related context of compact group actions on projective spaces, a basic theorem of Mumford, Sternberg, and Guillemin [16; 24, p. 158] asserts that the symplectic quotient $\mathscr{M}$ of $\mathscr{A}$ by $\mathscr{G}$ is naturally diffeomorphic to the quotient, in the sense of algebraic geometry, of $\mathscr{A}$ by $\mathscr{G}_c$. Since the latter receives a complex structure as a holomorphic quotient, $\mathscr{M}$ receives one also.

Those results about group actions on projective spaces carry over almost directly to our problem of certain types of group actions on affine spaces, using the fact that subgroups of the affine symplectic group obeying our hypotheses are actually extensions of compact groups by abelian ones. We will not develop this explicitly as actually the properties of the geometry of $\mathscr{M}$ that we need can be seen directly by local considerations near $F^{-1}(0)$, without appeal to the "global" results of geometric invariant theory. For instance, let us give a direct description of an almost complex structure $\tilde{J}$ obtained on $\mathscr{M}$ which coincides with the one given by its identification with $\mathscr{A}/\mathscr{G}_c$ when geometric invariant theory holds. (By the methods of §3a below, this almost complex structure can be shown to be integrable without reference to geometric invariant theory.) Let $\mathbf{g}_c$ be the complexification of the Lie algebra $\mathbf{g}$. The action of $\mathscr{G}_c$ is determined by the action of $\mathscr{G}$ and $\mathbf{g}_c$. Since we want it to be holomorphic in $\mathscr{G}_c$, the complex Lie algebra action $T_c : \mathbf{g}_c \to \mathrm{Vect}(\mathscr{A})$ must be

$$T_c(L_a) = T_a, \qquad T_c(iL_a) = JT_a. \tag{1.53}$$

At every $A \in F^{-1}(0)$, we have the following inclusion of spaces:

$$\begin{array}{ccc} TF^{-1}(0) & \subset & T\mathscr{A} \\ \cup & & \cup \\ T(\mathbf{g}) & \subset & T_c(\mathbf{g}_c) \end{array} \tag{1.54}$$

So we have the map

$$T_{\tilde{A}}\mathscr{M} \cong [T_{F^{-1}(0)}/T(\mathbf{g})]_A \to T\mathscr{A}/T_c(\mathbf{g}_c). \tag{1.55}$$

One can show that this is an isomorphism by simple dimension counting. As a quotient of complex vector spaces, the right-hand side of (1.55) receives a complex structure. Therefore, under the identification (1.55), $T_{\tilde{A}}\mathscr{M}$ receives a complex structure. (By $\mathscr{G}$-invariance, the choice of a point A in the orbit above $\tilde{A}$ is immaterial.) The fact that (1.55) is an isomorphism is just the infinitesimal version of the statement that $\mathcal{M} \cong \mathcal{A}/\mathcal{G}_c$. This isomorphism may also be proved using the Hodge theory description that we will present in §3. For instance, surjectivity of (1.55) follows since the representative of shortest length of any vector in $T\mathcal{A}/T_c(\mathbf{g}_c)$ actually lies in $TF^{-1}(0)$. This argument is just the infinitesimal version of the proof that $\mathcal{M} \cong \mathcal{A}/\mathcal{G}_c$, which uses a distance function to choose preferred elements on the $\mathcal{G}_c$ orbits.

Geometric invariant theory also constructs a holomorphic line bundle $\tilde{\mathcal{L}}$ over $\mathcal{A}/\mathcal{G}_c$, such that, if $\pi : \mathcal{A} \to \mathcal{A}/\mathcal{G}_c$ is the natural projection, then $\mathcal{L} = \pi^*(\tilde{\mathcal{L}})$. Moreover, $\tilde{\mathcal{L}}$ has the property that

$$H^0(\mathcal{A}, \mathcal{L})^G = H^0(\mathcal{A}/\mathcal{G}_c, \tilde{\mathcal{L}}). \tag{1.56}$$

This equation is a holomorphic analog of (1.50). Under the identification of $\mathcal{M}$ with $\mathcal{A}/\mathcal{G}_c$, the two definitions of $\tilde{\mathcal{L}}$ agree.

Connection for quantization of $\mathcal{M}$. We let $\tilde{\mathcal{H}}_{\mathrm{pr}}$ be the trivial prequantum bundle over $\mathcal{T}$ whose fibers are the sections of $\tilde{\mathcal{L}}$. We let $\tilde{\mathcal{H}}_Q$ be the bundle whose fibers consist of holomorphic sections of $\tilde{\mathcal{L}}$. The quantum bundle $\tilde{\mathcal{H}}_Q$ that arises in quantizing $\mathcal{M}$ may be identified with the $\mathcal{G}$-invariant subbundle of the quantum bundle $\mathcal{H}_Q$ that entered in our discussion of the quantization of the affine space $\mathcal{A}$:

$$\tilde{\mathcal{H}}_Q|_J = H^0_{L^2}\begin{pmatrix} \tilde{\mathcal{L}} \\ \downarrow \\ \mathcal{M} \end{pmatrix} \cong H^0_{L^2}\begin{pmatrix} \mathcal{L} \\ \downarrow \\ \mathcal{A} \end{pmatrix}^{\mathcal{G}_c} = H^0_{L^2}\begin{pmatrix} \mathcal{L} \\ \downarrow \\ \mathcal{A} \end{pmatrix}^{\mathcal{G}} = (\mathcal{H}_Q|_J)^{\mathcal{G}} \tag{1.57}$$

The second to last equality in (1.57) is due to the fact that, for holomorphic sections, $\mathcal{G}_c$-invariance is equivalent to $\mathcal{G}$-invariance. The $\mathcal{G}$-action on $\mathcal{H}_Q$ is just the prequantum action. It is invariant under parallel transport by $\delta^{\mathcal{H}_Q}$.

Thus we may identify $\tilde{\mathcal{H}}_Q$ with a subbundle of $\mathcal{H}_Q$ which is preserved under parallel transport by $\delta^{\mathcal{H}_Q}$. Therefore $\delta^{\mathcal{H}_Q}$ restricts to the desired projectively flat connection $\delta^{\tilde{\mathcal{H}}_Q}$ on the subbundle $\tilde{\mathcal{H}}_Q$. Of course, the Hermitian structure of $\tilde{\mathcal{H}}_Q$ is the one it inherits as a subbundle of $\mathcal{H}_Q$.

In finite dimensions, this is a complete description of the projectively flat connection on $\tilde{\mathcal{H}}_Q$; there is no need to say more. However, even in finite dimensions, one obtains a better understanding of the projectively flat connection on $\tilde{\mathcal{H}}_Q$ by describing it as much as possible in terms of the intrinsic geometry of $\mathcal{M}$. Moreover, the main application that we envision is to a gauge theory problem in which $\mathcal{A}$ and $\mathcal{G}$ are infinite dimensional, though the symplectic quotient $\mathcal{M}$ is finite dimensional. In this situation, the "upstairs" quantum bundle $\mathscr{H}_Q$ is difficult to define rigorously, though the "downstairs" bundle $\tilde{\mathscr{H}}_Q$ is certainly well defined. Under these conditions, we cannot simply define a connection on $\tilde{\mathscr{H}}_Q$ by restricting the connection on $\mathscr{H}_Q$ to the $\mathscr{G}$-invariant subspace. At best, we can construct a "downstairs" connection on $\tilde{\mathscr{H}}_Q$ by imitating the formulas that one would obtain if $\mathscr{H}_Q$, with a projectively flat connection of the standard form, did exist. Lacking a suitable construction of $\mathscr{H}_Q$, one must then check *ex post facto* that the connection that is constructed on $\tilde{\mathscr{H}}_Q$ has the correct properties. This is the program that we will pursue in this paper.

The desired connection on $\tilde{\mathscr{H}}_Q$ should have the following key properties. The connection form should be a second order differential operator on $\mathscr{M}$, since the connection upstairs has this property and since a second-order differential operator, restricted to act on $\mathscr{G}_c$-invariant functions, will push down to a second-order differential operator. The connection should be projectively flat. And it should be unitary.

In the gauge theory problem, we will be able to understand the first two of these properties. In fact, we will see that there is a natural connection on $\tilde{\mathscr{H}}_Q$ that is given by a second-order differential operator, and that this connection is projectively flat. Most of these properties (except for vanishing of the $(2, 0)$ part of the curvature) can be understood in terms of the local differential geometry of $\mathscr{M}$. Unitarity appears more difficult and perhaps can only be understood by referring back to the underlying infinite-dimensional affine space $\mathscr{A}$. It is even conceivable that a construction of the "upstairs" bundle $\mathscr{H}_Q$ is possible and would be the best approach along the lines of gauge theory to understanding the unitarity of the induced connection on the $\mathscr{G}_c$-invariant subbundle.

1c. Theta functions. As an example of the above ideas, we consider the case in which $\mathscr{A}$ is a $2n$-dimensional real vector space, with affine symplectic form ω and prequantum line bundle $\mathscr{L}$, and $\mathscr{G}$ is the discrete group of translations by a lattice Λ in $\mathscr{A}$ whose action on $\mathscr{A}$ lifts to an action on $\mathscr{L}$. Picking such a lift, and taking the Λ-invariant sections of $\mathscr{L}$, we get a prequantum bundle $\tilde{\mathscr{L}}$ over the torus $\mathscr{M} = \mathscr{A}/\Lambda$.

To quantize $\mathscr{M}$, we pick an affine complex structure J on $\mathscr{A}$ which defines a Kähler polarization; it descends to a complex structure $\tilde{J}$ on $\mathscr{M}$. The existence of the prequantum line bundle $\tilde{\mathscr{L}}$, with curvature of type $(1, 1)$, means that the complex torus $\mathscr{M}$ is actually a polarized abelian variety. The Hilbert space $\tilde{\mathscr{H}}_Q|_J = H^0(\mathscr{M}, \tilde{\mathscr{L}})$ serves as a quantization of $\mathscr{M}$.

This Hilbert space $\tilde{\mathscr{H}}_Q|_J$ is, in classical terminology, the space of theta functions for the polarized abelian variety $(\mathscr{M}, \tilde{\mathscr{L}})$.

Classically, one then goes on to consider the behavior as J varies in the Siegel upper half plane Ω, which parametrizes the affine Kähler polarizations of $\mathscr{A}$. Thus, the quantum Hilbert spaces $\tilde{\mathscr{H}}_Q|_J$, as J varies, fit together into a holomorphic bundle $\tilde{\mathscr{H}}_Q$ over Ω.

Since the work of Jacobi, it has been known that it is convenient to fix the theta functions, in their dependence on J, to obey a certain "heat equation". While it is well known that the theta functions, for fixed J, have a conceptual description as holomorphic sections of the line bundle $\tilde{\mathscr{L}}$, the conceptual origin of the heat equation which fixes the dependence on J is much less well known. In fact, this heat equation is most naturally understood in terms of the concepts that we have introduced above. The heat equation is simply the projectively flat connection $\delta^{\tilde{\mathscr{H}}_Q}$ on the quantum bundle $\tilde{\mathscr{H}}_Q$ over Ω which expresses the fact that, up to the usual projective ambiguity, the quantization of $\mathscr{M}$ is canonically independent of the choice of Kähler polarization $\tilde{J}$. The usual theta functions, which obey the heat equations, are projectively parallel sections of $\tilde{\mathscr{H}}_Q$. The fact that they obey the heat equation means that the quantum state that they represent is independent of J.

Actually, the connection $\delta^{\tilde{\mathscr{H}}_Q}$ as we have defined it in equation (1.34) is only projectively flat. The central curvature of this connection can be removed by twisting the bundle $\tilde{\mathscr{H}}$ by a suitable line bundle over Ω (with a connection whose curvature is minus that of $\delta^{\tilde{\mathscr{H}}_Q}$). The heat equation as usually formulated differs from $\delta^{\tilde{\mathscr{H}}_Q}$ by such a twisting. In the literature on geometric quantization, the twisting to remove the central curvature is called the "metaplectic correction". We have not incorporated this twisting in this paper because it cannot be naturally carried out in the gauge theory problem of interest.

In the rest of this subsection, we shall work out the details of the relation of the heat equation to the connection $\delta^{\tilde{\mathscr{H}}_Q}$. These details are not needed in the rest of the paper and can be omitted without loss.

Prequantization of $\mathscr{M}$. A prequantum Hermitian line bundle $\tilde{\mathscr{L}}$ on $\mathscr{M}$ with connection with curvature $-i\omega$ is, up to isomorphism, the quotient of $\mathscr{A} \times C$ under the identifications

$$(A, v) \sim (A+\lambda, e_\lambda(A)v),$$

where the "multipliers" e_λ can be taken to be of the form

$$e_\lambda(A) = \epsilon(\lambda)\exp(-\tfrac{i}{2}\omega(A - A_0, \lambda)). \tag{1.58}$$

Here A_0 is a point in $\mathscr{A}/\Lambda$ and may be considered as parametrizing the possible flat line bundles over $\mathscr{M}$, and $\epsilon(\lambda) \in \{\pm 1\}$ satisfies

$$\epsilon(\lambda_1 + \lambda_2) = \epsilon(\lambda_1)\epsilon(\lambda_2)(-1)^{\omega(\lambda_1, \lambda_2)/2\pi}. \tag{1.59}$$

A section of $\tilde{\mathscr{L}}$ is a function $s : \mathscr{A} \to C$ satisfying

$$s(A + \lambda) = e_\lambda(A)s(A). \tag{1.60}$$

The connection and metric can be taken to be

$$\nabla_i s(A) = \left(\frac{\partial}{\partial A^i} + \frac{i}{2}\omega_{ij}A^j\right) s(A), \qquad ||s||(A) = |s(A)|^2. \tag{1.61}$$

Family of complex structures on $\mathscr{M}$. The Siegel upper half-space Ω of complex, symmetric $n \times n$ matrices Z with positive imaginary part parametrizes affine Kähler polarizations of $\mathscr{A}$ and $\mathscr{M}$. For Z in Ω we may define the complex structure J_Z on $\mathscr{A}$ as follows. Fix an integral basis λ_i for Λ so that in terms of the dual coordinates $\{x_i\}$ on $\mathscr{A}$,

$$\frac{\omega}{2\pi} = \sum_i \delta_i dx^i \wedge dx^{i+n}. \tag{1.62}$$

Such a basis always exists (see [15]). (The δ_i are nonzero integers called elementary divisors; they depend on the choice of Λ.) The complex structure J_Z is defined by saying that the functions

$$A^{\underline{i}} = \frac{1}{2\pi}\left(\delta_i x^i + \sum_j Z_{ij}x^{j+n}\right), \qquad i = 1 \dots n, \tag{1.63}$$

are holomorphic. In terms of these,

$$\omega = \pi i \sum_{ij} dA^{\underline{i}}\, (\operatorname{Im} Z)^{-1}_{ij}\, dA^{\bar{j}}. \tag{1.64}$$

The map $Z \mapsto J_Z$ is a holomorphic map, which may be shown to map onto $\mathscr{T}$.

We easily compute

$$(\delta^{(1,0)} J_Z)^{\underline{i}}{}_{\bar{j}} = -((\delta^{(1,0)} Z)(\operatorname{Im} Z)^{-1})_{ij}. \tag{1.65}$$

The bundle $\tilde{\mathscr{H}}_Q$ *and the connection* $\delta^{\tilde{\mathscr{H}}_Q}$. The quantum Hilbert space $\tilde{\mathscr{H}}_Q|_{J_Z}$ is the space of holomorphic sections of $\tilde{\mathscr{L}}$ and is thus identified with functions s satisfying (1.60) and

$$0 = \nabla_{\bar{i}} s(Z, A) = \left(\frac{\partial}{\partial A^{\bar{i}}} + \frac{\pi}{2}(\operatorname{Im} Z)^{-1}_{ij} A^{\underline{j}}\right) s(Z, A). \tag{1.66}$$

As Z varies, such functions correspond to sections of the Hilbert bundle $\tilde{\mathscr{H}}_Q \to \mathscr{T}$ (pulled back to Ω).

It is not hard to write down explicitly the action of the connection $\delta^{\tilde{\mathscr{H}}_Q}$ on sections of $\tilde{\mathscr{H}}_Q$ realized in this way. Recall that

$$(1.67) \qquad \delta^{\tilde{\mathscr{H}}_Q} = \delta - M^{\underline{i}\underline{j}}\nabla_{\underline{i}}\nabla_{\underline{j}}, \qquad M^{\underline{i}\underline{j}} = -\tfrac{1}{4}(\delta J\omega^{-1})^{\underline{i}\underline{j}}.$$

(1.61) gives the actions of the covariant derivatives $\nabla_{\underline{i}}$ on s; and (1.64) and (1.65) give expressions for ω and $\delta^{(1,0)}J$. We need an expression for the action of δ on s. For any function $s(Z, A)$, let $\delta^Z s(Z, A)$ denote the exterior derivative in the Z directions when A is considered independent. It is given by the formula

$$(1.68) \quad \delta = \left[\delta^Z + \frac{1}{2i}(A^{\underline{i}} - A^{\bar{i}})\left(((\delta^{(1,0)}Z)(\operatorname{Im} Z)^{-1})_{ij}\frac{\partial}{\partial A^{\underline{j}}} + ((\delta^{(0,1)}Z)(\operatorname{Im} Z)^{-1})_{ij}\frac{\partial}{\partial A^{\bar{j}}}\right)\right],$$

where the second line takes into account the dependence of the coordinates $A^{\underline{i}}$ and $A^{\bar{i}}$ on Z. Substituting this expression and the formulas for $\nabla_{\underline{i}}$, ω, and δJ into (1.67) gives a formula for $\delta^{\tilde{\mathscr{H}}_Q}$ acting on sections of $\tilde{\mathscr{H}}_Q$ realized as functions s satisfying (1.60).

A more convenient formula, from the point of view of making contact with the traditional expressions for theta functions, however, is obtained by changing the trivialization of $\mathscr{L} = \mathscr{A} \times C$ so that holomorphic sections are represented by holomorphic functions of Z and A. Such a change in trivialization corresponds to a function $g : \Omega \times \mathscr{A} \to \mathbb{C}$ satisfying

$$(1.69) \qquad g\nabla_{\bar{i}}g^{-1} = \frac{\partial}{\partial A^{\bar{i}}} \quad \text{and} \quad g\delta^{(0,1)}g^{-1} = \delta^{Z(0,1)}.$$

To obtain the usual theta functions, we take

$$(1.70) \qquad g(Z, A) = \exp\left(\frac{\pi}{2}A^{\underline{i}}(\operatorname{Im} Z)^{-1}_{ij}(A^{\bar{j}} - A^{\underline{j}})\right).$$

If we now write $\theta(Z, A) = s(Z, A)g(Z, A)$, then the conditions that $s(Z, A)$ represent a holomorphic section of $\tilde{\mathscr{H}}_Q$ are that θ is holomorphic as a function of Z and A and has the periodicities

$$(1.71) \qquad \begin{aligned} \theta(Z, A + \lambda_i) &= \epsilon(\lambda_i)\theta(Z, A) \\ \theta(Z, A + \lambda_{i+n}) &= \epsilon(\lambda_{i+n})\exp(-2\pi i A^{\underline{i}} - \pi i Z_{ii})\theta(Z, A). \end{aligned}$$

The classical theta functions satisfy these conditions. The choice of ϵ is called the theta characteristic.

Acting on $\theta(Z, A)$, the holomorphic derivatives become

$$(1.72)\quad \begin{aligned} \nabla_{\underline{i}}\theta(Z, A) &= \left(\frac{\partial}{\partial A^{\underline{i}}} - \pi(\operatorname{Im} Z)^{-1}_{ij}(A^{\bar{j}} - A^{\underline{j}})\right)\theta(Z, A) \\ \delta^{(1,0)}\theta(Z, A) &= \Big[\delta^{Z(1,0)} + ((\delta^{(1,0)}Z)(\operatorname{Im} Z)^{-1})_{ij}(A^{\underline{j}} - A^{\bar{j}})\frac{\partial}{\partial A^{\underline{i}}} \\ &\quad + \frac{\pi}{4i}((\operatorname{Im} Z)^{-1}(\delta^{(1,0)}Z)(\operatorname{Im} Z)^{-1})_{ij} \\ &\quad \times (A^{\underline{i}} - A^{\bar{i}})(A^{\underline{j}} - A^{\bar{j}})\Big]\,\theta(Z, A). \end{aligned}$$

Combining the equations (1.64), (1.65), and (1.72), we find after a short calculation

$$(1.73)\quad \delta^{\bar{\mathscr{H}}_Q} = \delta^{\bar{\mathscr{H}}_Q'} - \frac{i}{4}\operatorname{Tr}(\operatorname{Im} Z^{-1}\delta^{(1,0)}Z),$$

where

$$(1.74)\quad \delta^{\bar{\mathscr{H}}_Q'(1,0)}\theta(Z, A) = \left[\delta^{Z(1,0)} - \frac{1}{4\pi i}(\delta^{(1,0)}Z)_{ij}\frac{\partial}{\partial A^i}\frac{\partial}{\partial A^j}\right]\theta(Z, A).$$

The modified connection $\delta^{\bar{\mathscr{H}}_Q'}$ has zero curvature. The equation $\delta^{\bar{\mathscr{H}}_Q'(1,0)}\theta(Z, A) = 0$ is the heat equation satisfied by the classical theta functions.

It may be shown that the modified connection $\delta^{\bar{\mathscr{H}}_Q'}$ is that obtained when account is taken of the "metaplectic correction". Thus the dependence of the classical theta functions on Z is naturally interpreted from this point of view as the statement that as Z varies, the theta functions $\theta(Z, \cdot) \in \bar{\mathscr{H}}_Q|_Z$ represent the same quantum state.

2. The gauge theory problem

In this section we will describe the concrete problem that actually motivated the investigation in this paper. It is the problem of quantizing the moduli space $\mathscr{M}$ of flat connections on a two dimensional surface Σ (of genus g, oriented, connected, compact, and without boundary). This moduli space can be regarded (as shown in [3]) as the symplectic quotient of an underlying infinite dimensional affine space by the action of the gauge group. Our goal is to explain concretely how this viewpoint leads to a projectively flat connection that makes possible quantization of

$\mathcal{M}$. In this section, we will aim for simplicity rather than precision and rigor. The precision and rigor will be achieved in later sections (which are independent of this one). Our goal in this section is to explain as directly as possible and without any unnecessary machinery the precise definition of the projectively flat connection that is used for quantization of $\mathcal{M}$, for the benefit of readers who may have use for the formulas. Also, we will explain in a language that should be familiar to physicists how one sees from this $(2+1)$-dimensional point of view a subtlety that is well known in $(1+1)$-dimensions, namely the replacement in many formulas of the "level" k by $k+h$, with h being the dual Coxeter number of the gauge group.

Preliminaries. Let G be a compact Lie group, which for convenience we take to be simple. The simple Lie algebra $\mathrm{Lie}(G)$ admits an invariant positive definite Killing form $(\, , \,)$, unique up to multiplication by a positive number. If F is the curvature of a universal G-bundle over the classifying space BG, a choice of $(\, , \,)$ enables us to define an element $\lambda = (F, \wedge F)$ of $H^4(BG, \mathbb{R})$. We normalize $(\, , \,)$ so that $\lambda/2\pi$ is a de Rham representative for a generator of $H^4(BG, \mathbb{Z}) \cong \mathbb{Z}$. This basic inner product $(\, , \,)$ is defined in down to earth terms in the appendix.

Let E be a principal G-bundle on the surface Σ. Let A be a connection on E. Locally, after picking a trivialization of E, A can be expanded

$$A = \sum A^a \cdot T_a, \tag{2.1}$$

where T_a is a basis of the Lie algebra $\mathrm{Lie}(G)$ of G. We can take T_a to be an orthonormal basis in the sense that $(T_a, T_b) = \delta_{ab}$. In this basis the Lie algebra $\mathrm{Lie}(G)$ may be described explicitly in terms of the structure constants $f_{ab}{}^c$:

$$[T_a, T_b] = f_{ab}{}^c T_c. \tag{2.2}$$

Defining $f_{abc} = f_{ab}{}^d \delta_{cd}$, invariance of $(\, , \,)$ implies that f_{abc} is completely antisymmetric.

One has

$$f_{ab}{}^c f_{cd}{}^b = 2\delta_{ad} \cdot h, \tag{2.3}$$

where h is the dual Coxeter number of G (as defined in the appendix).

Let $\mathcal{A}$ be the space of smooth connections on E. $\mathcal{A}$ is an affine space; its tangent space $T\mathcal{A}$ consists of one forms on Σ with values in $\mathrm{ad}(E)$. $\mathcal{A}$ has a natural symplectic structure, determined by the symplectic form

$$\omega_0 = \frac{1}{4\pi} \int_\Sigma \delta A^a \wedge \delta A^b \, \delta_{ab}. \tag{2.4}$$

The symplectic form that we will actually use in quantization is

$$\omega = k \cdot \omega_0, \tag{2.5}$$

with k a positive integer called the "level".

After picking local coordinates on Σ, the connection d_A can be described by the explicit formula for covariant derivatives

$$D_\mu \phi = (\partial_\mu + A_\mu{}^a T_a)\phi, \tag{2.6}$$

with ϕ denoting a section of any associated bundle to E. And the curvature form is

$$F_{\mu\nu}{}^a = \partial_\mu A_\nu{}^a - \partial_\nu A_\mu{}^a + f_{cd}{}^a A_\mu{}^c A_\nu{}^d. \tag{2.7}$$

The "gauge group" $\mathcal{G}$ is the group of automorphisms of E as a principal bundle. An element g of the gauge group transforms the connection d_A by

$$d_A \to g \cdot d_A \cdot g^{-1}. \tag{2.8}$$

$\mathcal{G}$ is an infinite-dimensional Lie group whose Lie algebra $\mathbf{g}$ consists of the smooth sections of $\mathrm{ad}(E)$. The action of $\mathbf{g}$ on $\mathcal{A}$ is described by the map

$$\epsilon \to -d_A \epsilon \tag{2.9}$$

from $\Gamma(\mathrm{ad}(E))$ to $T_A(\mathcal{A})$.

The action of the group $\mathcal{G}$ on the space $\mathcal{A}$ of connections preserves the symplectic structure ω. The moment map $F : \mathcal{A} \to \mathbf{g}^\vee$ is the map which takes a connection A to its curvature. The curvature F is a two-form with values in $\mathrm{ad}(E)$ and is identified with an element of $\mathbf{g}^\vee$ by the pairing

$$< F, \epsilon > = \frac{k}{4\pi} \int_\Sigma (\epsilon, F) \tag{2.10}$$

for $\epsilon \in \Gamma(\mathrm{ad}(E))$.

The zeros of the moment map $F^{-1}(0)$ thus consist of flat connections, and the quotient $\mathcal{M} = F^{-1}(0)/\mathcal{G}$ is therefore the moduli space of flat connections on E, up to gauge transformation. If G is connected and simply connected, $\mathcal{M}$ is simply the moduli space of flat G bundles over Σ. If, however, G is not connected and simply connected, there may be several topological types of flat G bundles on Σ, and $\mathcal{M}$ is the moduli space of flat bundles with the topological type of E.

The general arguments about symplectic quotients apply in this situation, so that the symplectic structure ω_0 on $\mathcal{A}$ descends to a symplectic structure on $\mathcal{M}$, which we will call $\tilde{\omega}_0$. Our normalization convention on $(\, , \,)$ ensures that $\tilde{\omega}_0/2\pi$ represents an integral element of $H^2(\mathcal{M}, \mathbb{R})$, so that $\tilde{\omega}_0$ is at least prequantizable.

Holomorphic interpretation. Our goal is to quantize $\mathcal{M}$ with the symplectic structure ω_0 and more generally with the symplectic structure $\tilde{\omega} = k\tilde{\omega}_0$.

One of the important ingredients will be a construction of a suitable complex structure on $\mathcal{A}$. To do so, we pick a complex structure J on Σ (such that the orientation on Σ determined by J is the given one). This choice induces a complex structure $J_{\mathcal{A}}$ on $\mathcal{A}$, as follows. The tangent space $T\mathcal{A}$ consists of one forms on Σ with values in $\mathrm{ad}(E)$. Given a complex structure J on Σ, we define

$$J_{\mathcal{A}}\delta A = -J\,\delta A, \qquad \delta A \in T\mathcal{A}. \tag{2.11}$$

Relative to this complex structure

$$T\mathcal{A} = T^{(1,0)}\mathcal{A} \oplus T^{(0,1)}\mathcal{A}, \tag{2.12}$$

where $T^{(1,0)}\mathcal{A}$ and $T^{(0,1)}\mathcal{A}$ consist respectively of $(0,1)$-forms and $(1,0)$-forms on Σ with values in $\mathrm{ad}(E)$. (This is opposite to the choice of complex structure on $\mathcal{A}$ which appears frequently in the physics literature in which the holomorphic directions are represented by holomorphic one-forms on Σ. The above choice, however, is more natural since it is the antiholomorphic one-forms which couple to the $\overline{\partial}$ operator on Σ, and it is this operator, which defines a complex structure on the bundle E, which we want to vary holomorphically as a function of the complex structures on Σ and $\mathcal{A}$.) It is evident that with the choice (2.12) the symplectic form ω on $\mathcal{A}$ is positive and of type $(1,1)$.

By analogy with the discussion in §1 of group actions on finite-dimensional affine spaces, one might expect that once the complex structure $J_{\mathcal{A}}$ is picked, the action of the gauge group $\mathcal{G}$ can be analytically continued to an action of the complexified gauge group $\mathcal{G}_c$ (which, in local coordinates, consists of smooth maps of Σ to G_c, the complexification of G). It is easy to see that this is so. Once the complex structure J is picked on Σ, the connection d_A can be decomposed as

$$d_A = \partial_A + \overline{\partial}_A, \tag{2.13}$$

where ∂_A and $\overline{\partial}_A$ are the $(1,0)$ and $(0,1)$ pieces of the connection, respectively. The $\mathcal{G}_c$ action on connections is then determined by the formula

$$\overline{\partial}_A \to g \cdot \overline{\partial}_A \cdot g^{-1}. \tag{2.14}$$

It is evident that this action is holomorphic. (2.14) implies the complex conjugate formula

$$\partial_A \to \overline{g} \cdot \partial_A \cdot \overline{g}^{-1}, \tag{2.15}$$

with $\overline{g}$ the complex conjugate of g.

Equation (2.14) has the following interpretation. For dimensional reasons, the $(0, 2)$ part of the curvature of any connection on a Riemann surface Σ vanishes. Therefore, for any connection A, the $\overline{\partial}_A$ operator gives a holomorphic structure to the principal G_c-bundle E_c (E_c is the complexification of E). The holomorphic structures determined by two such operators $\overline{\partial}_A$ and $\overline{\partial}_{A'}$ are equivalent if and only $\overline{\partial}_A$ and $\overline{\partial}_{A'}$ are conjugate by a transformation of the kind (2.14), that is, if and only if they are on the same orbit of the action of $\mathcal{G}_c$ on $\mathcal{A}$. Therefore, the set $\mathcal{A}/\mathcal{G}_c$ can be identified with the set $\mathcal{M}_J^{(0)}$ of equivalence classes of holomorphic structures on E_c.

Under a suitable topological restriction, for instance, if $G = SO(3)$ and E is an $SO(3)$ bundle over Σ with nonzero second Stieffel-Whitney class, $\mathcal{G}_c$ acts freely on $\mathcal{A}$. In this case, we will simply refer to $\mathcal{M}_J^{(0)}$ as $\mathcal{M}_J$; it is the moduli space of holomorphic G_c bundles over Σ, of specified topological type. The subscript in $\mathcal{M}_J$ is meant to emphasize that these bundles are holomorphic in the complex structure J. In general, reducible connections correspond to singularities in the quotient $\mathcal{A}/\mathcal{G}_c$; in this case, instead of the naive set theoretic quotient $\mathcal{A}/\mathcal{G}_c$, one should take the quotient in the sense of geometric invariant theory. Doing so, one gets the moduli space $\mathcal{M}_J$ of semistable G_c bundles on Σ, of a fixed topological type.

For finite-dimensional affine spaces, we know that the symplectic quotient by a compact group can be identified with the ordinary quotient by the complexified group. Does such a result hold for the action of the gauge group $\mathcal{G}$ on the infinite-dimensional affine space $\mathcal{A}$? The symplectic quotient of $\mathcal{A}$ by $\mathcal{G}$ is the moduli space $\mathcal{M}$ of flat connections on E; the ordinary quotient of $\mathcal{M}$ by $\mathcal{G}_c$ gives the moduli space $\mathcal{M}_J$ of holomorphic structures on E_c. In fact, there is an obvious map $i : \mathcal{M} \to \mathcal{M}_J$ coming from the fact that any flat structure on E determines a holomorphic structure on E_c. Using Hodge theory, it is easy to see that the map i induces an isomorphism of the tangent spaces of $\mathcal{M}$ and $\mathcal{M}_J$. Indeed, $T\mathcal{M} = H^1(\Sigma, \mathrm{ad}(E_A))$. (Here $H^*(\Sigma, \mathrm{ad}(E_A))$ denotes de Rham cohomology of Σ with values in the flat bundle $\mathrm{ad}(E_A)$.) According to the

Hodge decomposition, the complexification of $H^1(\Sigma, \mathrm{ad}(E_A))$ is

$$H^1_{\mathbb{C}}(\Sigma, \mathrm{ad}(E_A)) = H^{(0,1)}(\Sigma, \mathrm{ad}(E_A)) \oplus H^{(1,0)}(\Sigma, \mathrm{ad}(E_A)), \tag{2.16}$$

and on the right-hand side we recognize the $(1,0)$ and $(0,1)$ parts of the complexified tangent bundle of $\mathcal{M}_J$.

Actually, it is a fundamental theorem of Narasimhan and Seshadri that the map i is an isomorphism between $\mathcal{M}$ and $\mathcal{M}_J$. Just as in our discussion of the comparison between symplectic quotients and holomorphic quotients of finite-dimensional affine spaces, the symplectic form ω becomes a Kähler form on $\mathcal{M}_J$ under this isomorphism. In particular, ω is of type $(1,1)$, so any prequantum line bundle $\tilde{\mathcal{L}}$ over $\mathcal{M}$ becomes a holomorphic line bundle on $\mathcal{M}_J$ (and in fact taking the holomorphic sections of powers of $\tilde{\mathcal{L}}$ leads to an embedding of $\mathcal{M}_J$ in projective space).

Thus, the Narasimhan-Seshadri theorem gives us a situation similar to the situation for symplectic quotients of affine spaces in finite dimensions. As the complex structure J on Σ varies, the $\mathcal{M}_J$ vary as Kähler manifolds; but as symplectic manifolds they are canonically isomorphic to a fixed symplectic variety $\mathcal{M}$.

Since diffeomorphisms that are isotopic to the identity act trivially on $\mathcal{M}$, isotopic complex structures J and J' give the *same* Hodge decompositions (2.16), not just "equivalent" ones. Therefore, the complex structure of $\mathcal{M}_J$ depends on the complex structure J on Σ only up to isotopy.

The moduli space of complex structures on Σ up to isotopy is usually called the Teichmuller space of Σ; we will denote it as $\mathcal{T}$. A point $t \in \mathcal{T}$ does not determine a canonical complex structure on Σ (it determines one only up to isotopy). But in view of the above, the choice of t does determine a canonical complex structure J_t on the moduli space $\mathcal{M}$ of flat connections. The complex structure J_t varies holomorphically in t. Therefore, the product $\mathcal{M} \times \mathcal{T}$, regarded as a bundle over $\mathcal{T}$, gets a natural complex structure, with the fibers being isomorphic as symplectic manifolds but the complex structure of the fibers varying with t.

Prequantization and the action of the mapping class group. There are several rigorous approaches to constructing a prequantum line bundle $\tilde{\mathcal{L}}$ over $\mathcal{M}$—that is, a unitary line bundle with a connection of curvature $-i\tilde{\omega}$. Since for G a compact, semisimple Lie group, $b_1(\mathcal{M})$ vanishes and $H^1(\mathcal{M}, U(1))$ is a finite set, there are finitely many isomorphism classes of such prequantum line bundles.

Holomorphically, one can pick a complex structure J on Σ, and take $\tilde{\mathcal{L}}$ to be the determinant line bundle Det_J of the $\overline{\partial}_J$ operator coupled to the associated bundle $E(R)$ determined by some representation R of the gauge group. If this line bundle is endowed with the Ray-Singer-Quillen metric, then the main result of [27] shows that its curvature is $-il(R)\cdot\tilde{\omega}_0$, where $l(R)$ is defined by $-\mathrm{Tr}_R(T_aT_b) = l(R)\delta_{ab}$. Though this construction depends on a choice of complex structure on Σ, the line bundle Det_J, as a unitary line bundle with connection, is independent of the complex structure chosen since the space of complex structures is connected and (according to the remark at the end of the last paragraph) the set of isomorphism classes of prequantum bundles is a finite set. (The isomorphism among the Det_J as J varies can also be seen more explicitly by using the Quillen connection to define a parallel transport on the Det_J bundle as J varies.) Thus, if k is of the form $2c_2(R)$ for some not necessarily irreducible representation of G, the prequantum bundle can be defined as a determinant line bundle.

From the point of view of the present paper, it is more natural to construct the prequantum line bundle by pushing down a trivial prequantum bundle $\mathscr{L}$ from the underlying infinite-dimensional affine space $\mathscr{A}$. This can be done rigorously [28].[4]

Once E is fixed, there is some subgroup $\Gamma_{\Sigma,E}$ of the mapping class group of Σ consisting of diffeomorphisms ϕ that fix the topological type of E. $\Gamma_{\Sigma,E}$ has an evident action on $\mathscr{M}$ coming from the interpretation of the latter as a moduli space of representations of $\pi_1(\Sigma)$. The goal of the present paper is to construct an action of $\Gamma_{\Sigma,E}$ on the Hilbert spaces obtained by quantizing $\mathscr{M}$. For this aim, we lift the action of the mapping class group of Σ on $\mathscr{M}$ to an action (or at least a projective action) on the prequantum line bundle $\tilde{\mathscr{L}}$.

Actually, if the prequantum line bundle is unique up to isomorphism, which occurs if G is connected and simply connected in which case $H^1(\mathscr{M}, U(1)) = 0$, then at least a projective action of the mapping class group is automatic. Even if the prequantum line bundle is not unique up to isomorphism, on a prequantum line bundle constructed as a determinant line bundle one automatically gets a projective action of the mapping class group. (If ϕ is a diffeomorphism of Σ, and $\tilde{\mathscr{L}}$ has been constructed as Det_J for some J, then ϕ naturally maps $\det_J$ to $\det_{\phi J}$, which has a projective identification with $\det_J$ noted in the last paragraph.)

The construction of prequantum line bundles via pushdown is also a natural framework for constructing actions (not just projective actions) of the

[4]The authors of that paper consider explicitly the case of $G = SU(2)$, but it should be straightforward to generalize their constructions.

mapping class group. We will now sketch how this construction arises from the three-dimensional point of view. This will be discussed more precisely elsewhere. To start with, we choose an element of $H^4(BG, \mathbb{Z})$. This allows us to define an $\mathbb{R}/2\pi\mathbb{Z}$ valued "Chern-Simons" functional of connections on G bundles over three-manifolds with boundary, as discussed in [6]. The functional S obeys the factorization property that $e^{iS(M_0, E_0, A_0)} = e^{iS(M_1, E_1, A_1)} \cdot e^{iS(M_2, E_2, A_2)}$ if the three-manifold M_0 with bundle E_0 and connection A_0 is obtained by gluing M_1, E_1, A_1 to M_2, E_2, A_2. The gluing is accomplished by an identification, $\Phi : E_1|_{\Sigma_1} \to E_2|_{\Sigma_2}$, of the restriction of E_2 to some boundary component Σ_1 of M_1 with the restriction of E_2 to some boundary component Σ_2 of M_2. Now fix a bundle E over a surface Σ. Let A be an element of the space $\mathscr{A}$ of connections on E. For $i = 1, 2$, let (M_i, E_i, A_i) be obtained by crossing (Σ, E, A) with an interval. So for each automorphism Φ of E (not necessarily base preserving) we may form M_0, E_0, A_0 by gluing. We thus obtain a function $\rho(\Phi, A) = e^{iS(M_0, E_0, A_0)}$ from $\mathrm{Aut}(E) \times \mathscr{A}$ to $U(1)$. By factorization, ρ is a lift of the action of $\mathrm{Aut}(E)$ on $\mathscr{A}$ to the trivial line bundle over $\mathscr{A}$. By restricting to flat connections and factoring out by the normal subgroup $\mathrm{Aut}'(E)$ of $\mathrm{Aut}(E)$ consisting of automorphism which lift diffeomorphisms of Σ which are connected to the identity, we obtain the line bundle $\tilde{\mathscr{L}}$ over $\mathscr{M}$ with an action of the mapping class group $\Gamma_{\Sigma, E} = \mathrm{Aut}(E)/\mathrm{Aut}'(E)$.

Finally, we can introduce the action of the mapping class group $\Gamma_{\Sigma, E}$ on the quantum bundle $\tilde{\mathscr{H}}_Q$ over $\mathscr{T}$. The fiber $\tilde{\mathscr{H}}_Q|_t$ of $\tilde{\mathscr{H}}_Q$, over a point $t \in \mathscr{T}$, is simply $H^0(\mathscr{M}_{J_t}, \tilde{\mathscr{L}})$. Obviously, since the mapping class group has been seen to act on $\tilde{\mathscr{L}}$, its action on $\mathscr{T}$ lifts naturally to an action on $\tilde{\mathscr{H}}_Q$. Our goal is to construct a natural, projectively flat connection ∇ on the bundle $\tilde{\mathscr{H}}_Q \to \mathscr{T}$. Naturalness will mean in particular that ∇ is invariant under the action of $\Gamma_{\Sigma, E}$. A projectively flat connection on $\mathscr{T}$ that is $\Gamma_{\Sigma, E}$-invariant determines a projective representation of $\Gamma_{\Sigma, E}$. Thus, in this way we will obtain representations of the genus g mapping class groups. These representations are genus g counterparts of the Jones representations of the braid group.

The precise connection with Jones's work depends on the following. At least formally, one can generalize the constructions to give representations of the mapping class groups $\Gamma_{g,n}$ for a surface of genus g with n marked points $P_1, \cdots, P_n$. This is done by considering flat connections on $\Sigma - \bigcup_i P_i$ with prescribed monodromies around the P_i. Jones's representations would then correspond to some of the representations so obtained for $\Gamma_{0,n}$. For simplicity, we will only consider the case without marked points.

Construction of the connection. We will now describe how, formally, one can obtain the desired connection on the bundle $\tilde{\mathscr{H}}_Q \to \mathscr{T}$, by formally supposing that one has a quantization of the infinite-dimensional affine space $\mathscr{A}$, and "pushing down" the resulting formulas from $\mathscr{A}$ to $\mathscr{M}$.

To begin with, we must (formally) quantize $\mathscr{A}$. This is done by using the complex structure $J_{\mathscr{A}}$ on $\mathscr{A}$ that comes (as described above) from a choice of a complex structure J on Σ. The prequantum line bundle $\mathscr{L}$ over $\mathscr{A}$ is a unitary line bundle with a connection ∇ of curvature $-i\omega$. To describe this more explicitly, let z be a local complex coordinate on σ, and define $\delta/\delta A_z{}^a(z)$ and $\delta/\delta A_{\bar{z}}{}^a(z)$ by

$$\begin{aligned}\nabla_u \psi(A) &= \int_\Sigma d^2 z\, u_z^a \frac{\delta}{\delta A_z^a(z)} \psi(A), \\ \nabla_{\bar{u}} \psi(A) &= \int_\Sigma d^2 z\, \bar{u}_{\bar{z}}^a \frac{\delta}{\delta A_{\bar{z}}^a(z)} \psi(A)\end{aligned} \tag{2.17}$$

for u and $\bar{u}$ adjoint valued $(1,0)$ and $(0,1)$ forms on Σ and $d^2 z = i dz d\bar{z}$. (We will sometimes abbreviate $\int_\Sigma d^2 z$ as $\int_\Sigma$.) Then the connection ∇ is characterized by

$$\left[\frac{\delta}{\delta A_w{}^a(w)}, \frac{\delta}{\delta A_{\bar{z}}{}^b(z)}\right] = -i\frac{k}{4\pi}\delta_{ab}\delta_{z\bar{w}}(z, w), \tag{2.18}$$

along with

$$\left[\frac{\delta}{\delta A_z{}^a(z)}, \frac{\delta}{\delta A_w{}^b(w)}\right] = \left[\frac{\delta}{\delta A_{\bar{z}}{}^a(z)}, \frac{\delta}{\delta A_{\bar{w}}{}^b(w)}\right] = 0. \tag{2.19}$$

(In (2.18), $\delta_{z\bar{w}}(z, w) dz d\bar{w}$ represents the identity operator on $\Gamma(K \otimes \mathrm{ad}(E))$; that is $\int_{\Sigma_w} d\bar{w} dw \delta_{z\bar{w}} u_w = u_z$. We also have

$$\int_{\Sigma_z} d\bar{z} dz\, \bar{u}_{\bar{z}} \delta_{z\bar{w}} = \bar{u}_{\bar{w}}.)$$

The "upstairs" quantum Hilbert space $\mathscr{H}_Q|_J$ consists of holomorphic sections Ψ of the prequantum bundle, that is, sections obeying

$$\frac{\delta}{\delta A_z{}^a(z)} \Psi = 0. \tag{2.20}$$

What we actually wish to study is the object $\tilde{\mathscr{H}}_Q|_J = H^0(\mathscr{M}_J, \tilde{\mathscr{L}})$ introduced in the last subsection. The latter is perfectly well defined. Formally, this well-defined object should be the $\mathscr{G}_c$-invariant subspace of the larger space $\mathscr{H}_Q|_J$. At present, the latter is ill defined.

But proceeding formally, we will attempt to interpret $\tilde{\mathscr{H}}_Q|_J$ as the $\mathscr{G}_c$-invariant subspace of $\mathscr{H}_Q|_J$. Supposing for simplicity that $\mathscr{G}$ is connected, the $\mathscr{G}_c$-invariant subspace is the same as the subspace invariant under the Lie algebra $\mathbf{g}_c$ of $\mathscr{G}_c$. The condition for $\mathbf{g}_c$-invariance is

$$\left(-D_{\overline{z}}\frac{\delta}{\delta A_{\overline{z}}{}^{a}(z)}+\frac{k}{4\pi}\delta_{ab}F^{b}_{\overline{z}z}(z)\right)\Psi=0. \tag{2.21}$$

Here $F^a_{\overline{z}z}(z)$ is the curvature of the connection A, which enters because it is the moment map in the action of the gauge group on the space of connections. Also, $D_{\overline{z}}=\frac{\partial}{\partial\overline{z}}+A_{\overline{z}}$ is the $(0,1)$ component of the exterior derivative coupled to A.

Green's functions. To proceed further, it is convenient to introduce certain useful Green's functions that arise in differential geometry on the smooth surface Σ. In what follows, we will be working with flat connections on Σ—corresponding to zeros of the moment map for the $\mathscr{G}$-action on $\mathscr{A}$. For flat connections, the relevant Green's functions can all be expressed in terms of the Green's function of the Laplacian, or equivalently the operator $\overline{\partial}\partial$. For simplicity, we shall assume that this operator has no kernel. Let $\pi_i:\Sigma\times\Sigma\to\Sigma$, for $i=1,2$, be the projections on the first and second factors respectively. Let $E_i=\pi_i^*(E)$, for $i=1,2$. Similarly, let K be the canonical line bundle of Σ, regarded as a complex Riemann surface with complex structure J, and let $K_i=\pi_i^*(K)$.

The Green's function for the operator $\overline{\partial}\partial$ is a section ϕ of $\mathrm{ad}(E_1)\otimes\mathrm{ad}(E_2)^\vee$ over $\Sigma\times\Sigma-\Delta$ (Δ being the diagonal), such that

$$D_{\overline{z}}D_z\phi^a{}_b(z,w)=\delta^a{}_b\delta_{\overline{z}z}(z,w). \tag{2.22}$$

(Here $\delta_{\overline{z}z}(z,w)$ satisfies $\int_{\Sigma_w}d\overline{w}dw\delta_{\overline{z}z}v_{\overline{w}w}=v_{\overline{z}z}$.) It is convenient to also introduce

$$L_z{}^a{}_b(z,w)=D_z\phi^a{}_b(z,w) \tag{2.23}$$

and

$$\overline{L}_{\overline{z}}{}^a{}_b(z,w)=D_{\overline{z}}\phi^a{}_b(z,w). \tag{2.24}$$

They are sections, respectively, of $K_1\otimes\mathrm{ad}(E_1)\otimes\mathrm{ad}(E_2)^\vee$ and $\overline{K}_1\otimes\mathrm{ad}(E_1)\otimes\mathrm{ad}(E_2)^\vee$, over $\Sigma\times\Sigma-\Delta$, and obviously obey (for A a flat connection)

$$D_{\overline{z}}L_z{}^a{}_b(z,w)=-D_z\overline{L}_z{}^a{}_b(z,w)=\delta^a{}_b\delta_{\overline{z}z}(z,w). \tag{2.25}$$

On $H^0(K \otimes \mathrm{ad}(E))$, there is a natural Hermitian structure given by

$$|\lambda|^2 = \frac{1}{i}\int_\Sigma \overline{\lambda}^a \wedge \lambda^b \delta_{ab}. \tag{2.26}$$

Let $\lambda_{(i)}{}^a(z)$, $i = 1\ldots(g-1)\cdot \mathrm{Dim}(G)$, be an orthonormal basis for $H^0(K \otimes \mathrm{ad}(E))$, that is, an orthonormal basis of $\mathrm{ad}(E)$-valued $(1, 0)$ forms obeying

$$D_{\overline{z}}\lambda_{(i)z}{}^a(z) = 0. \tag{2.27}$$

Obviously, their complex conjugates $\overline{\lambda}_{(i)}{}^a$ are $\mathrm{ad}(E)$-valued $(0, 1)$-forms that furnish an orthonormal basis of solutions of the complex conjugate equation

$$D_z\overline{\lambda}_{(i)\overline{z}}{}^a(z) = 0. \tag{2.28}$$

Finally, we have

$$D_{\overline{w}}L_z{}^a{}_b(z, w) = -\delta^a{}_b\delta_{z\overline{w}}(z, w) + \frac{1}{i}\sum_i \lambda_{(i)z}{}^a(z)\overline{\lambda}^b_{(i)\overline{w}}(w), \tag{2.29}$$

together with the complex conjugate equation.

We can now re-express (2.21) in a form that is convenient for constructing the well-defined expressions on $\mathscr{M}_J$ that are formally associated with ill-defined expressions on $\mathscr{A}$. Let

$$\mathscr{D}_{(i)} = \nabla_{\overline{\lambda}_{(i)}} = \int_\Sigma \overline{\lambda}_{(i)\overline{z}}{}^a(z)\frac{\delta}{\delta A_{\overline{z}}{}^a(z)}. \tag{2.30}$$

And for future use, let

$$\overline{\mathscr{D}}_{(i)} = \nabla_{\lambda_{(i)}} = \int_\Sigma \lambda_{(i)z}{}^a(z)\frac{\delta}{\delta A_z{}^a(z)}. \tag{2.31}$$

Then

$$\frac{\delta}{\delta A_{\overline{z}}{}^a(z)} = i\int_{\Sigma_w} L_z{}^a{}_b(z, w)D_{\overline{w}}\frac{\delta}{\delta A_{\overline{w}}{}^b(w)} + \sum_i \lambda_{(i)z}{}^a(z)\mathscr{D}_{(i)}. \tag{2.32}$$

The symbol $\int_{\Sigma_w}$ is just an instruction to integrate over the w variable. (2.32) is proved by integrating D_w by parts and using (2.29).

At last, we learn that on a section Ψ of $\mathscr{H}_Q|_J$ that obeys (2.21), we can write

$$\frac{\delta}{\delta A_{\overline{z}}{}^a(z)} = \frac{k}{4\pi}\int_{\Sigma_w} d\overline{w}dw L_z{}^a{}_b(z, w)F^b_{\overline{w}w}(w) + \sum_i \lambda_{(i)z}{}^a(z)\mathscr{D}_{(i)}. \tag{2.33}$$

The point of this formula is that arbitrary derivatives with respect to $A_{\bar z}$, appearing on the left, are expressed in terms of derivatives $\mathscr{D}_{(i)}$ in finitely many directions that correspond exactly to the tangent directions to $\mathscr{M}$.

We also will require formulas for the change in the $\lambda_{(i)}{}^a(z)$ and their complex conjugates $\bar\lambda_{(i)}{}^a(z)$ under a change in the flat connection A. There is some arbitrariness here, since although the sum

$$\frac{1}{i}\sum_i \lambda_{(i)}{}^a(z)\bar\lambda_{(i)}{}^b(w) \tag{2.34}$$

is canonical—it represents the projection operator onto the kernel of $\bar\partial_A$ acting on one-forms—the individual $\lambda_{(i)}{}^a(z)$ are certainly not canonically defined. However, expanding around a particular flat connection A, it is convenient to choose the $\bar\lambda_{(i)}$ so that, to first order in δA,

$$\frac{\delta}{\delta A_{\bar z}{}^a(z)}\bar\lambda_{(i)\bar w}{}^b(w) = 0. \tag{2.35}$$

This choice is natural since $A_{\bar z}$ does not appear in the equation obeyed by the $\bar\lambda_{(i)}$ (in other words this equation depends antiholomorphically on the connection). Varying (2.28) with respect to the connection and requiring that the orthonormality of the $\lambda_{(i)}$ should be preserved then leads to

$$\frac{\delta}{\delta A_{\bar z}{}^a(z)}\lambda_{(i)w}{}^b(w) = -iL_w{}^b{}_d(w\,,\,z)\,f_{ac}{}^d\lambda_{(i)z}{}^c(z). \tag{2.36}$$

Construction of the connection. We are now in a position to construct the desired connection on the quantum bundle $\mathscr{H}_Q \to \mathscr{T}$. First, we work "upstairs" on $\mathscr{A}$. To define a complex structure on $\mathscr{A}$, we need an actual complex structure on Σ, not just one defined up to isotopy. Accordingly, we shall also work over the space of complex structures on Σ. We discuss below why, for the final answer, these complex structures need actually only be defined up to isotopy. The projectively flat connection on $\mathscr{H}_Q$ that governs quantization of $\mathscr{A}$ is formally

$$\delta^{\mathscr{H}_Q} = \delta - \frac{it}{4}\cdot\frac{4\pi}{k}\int_\Sigma \delta J_{\bar z}{}^z\,\frac{\delta}{\delta A_{\bar z}{}^a(z)}\frac{\delta}{\delta A_{\bar z}{}^a(z)}. \tag{2.37}$$

Given the formulas for the symplectic structure and complex structure of $\mathscr{A}$, (2.37) is an almost precise formal transcription of the basic formula—equation (1.34)—for quantization of an affine space. However, for finite-dimensional affine spaces, one would have $t = 1$, as we see in (1.34); in the present infinite-dimensional situation, it is essential, as we will see, to permit ourselves the freedom of taking $t \neq 1$.

We now wish to restrict $\delta^{\mathscr{H}_Q}$ to act on $\mathscr{G}_c$-invariant sections Ψ of $\tilde{\mathscr{H}}_Q$, that is, sections that obey (2.33). On such sections, we can use (2.33) to write

(2.38)
$$\delta^{\mathscr{H}_Q} = \delta - \frac{it\pi}{k}\int_\Sigma \delta J_{\bar{z}}{}^z \frac{\delta}{\delta A_{\bar{z}}{}^a(z)} \times \left(\frac{k}{4\pi}\int_{\Sigma_w} d\overline{w}dw L_z{}^a{}_b(z, w)F^b_{\overline{w}w}(w) + \sum_i \lambda_{(i)z}{}^a(z)\mathscr{D}_{(i)}\right).$$

We now want to move $\delta/\delta A_{\bar{z}}{}^a(z)$ to the right on (2.41), so that—acting on $\mathscr{G}_c$ invariant sections—we can use (2.33) again. At this point, however, it is convenient to make the following simplification. Our goal is to obtain a well-defined connection on sections of $\tilde{\mathscr{L}}$ over $\mathscr{M}\times\{J\}$ which are holomorphic in $\mathscr{M}$. Since we are working at flat connections, after moving $\delta/\delta A_{\bar{z}}{}^a(z)$ to the right in (2.38), we are entitled to set $F = 0$. This causes certain terms to vanish.

In moving $\delta/\delta A_{\bar{z}}{}^a$ to the right, we encounter a term

$$\frac{\delta}{\delta A_{\bar{z}}{}^a(z)}F^b_{\overline{w}w}(w) = -iD_w\left(\delta_a{}^b\delta_{z\overline{w}}(z, w)\right). \tag{2.39}$$

We also pick up a term

$$\frac{\delta}{\delta A_{\bar{z}}{}^a(z)}\lambda_{(i)z}{}^a(z) = -iL_z{}^a{}_d(z, z)\, f_{ac}{}^d\lambda_{(i)z}{}^c(z), \tag{2.40}$$

where we have blindly used (2.36); the meaning of this formal expression that involves the value of a Green's function on the diagonal will have to be discussed later. With the help of (2.39) and (2.40), one finds that upon moving $\delta/\delta A_{\bar{z}}{}^a(z)$ to the right and setting $F = 0$ one gets

$$\delta^{\mathscr{H}_Q} = \delta + \frac{t\pi}{k}\int_\Sigma\left(-\frac{k}{4\pi}\int_{\Sigma_w} d\overline{w}dw L_z{}^a{}_b(z, w)(D_w(\delta^b{}_a\delta(z, w))) - L_z{}^a{}_d(z, z)\, f_{ac}{}^d\lambda_{(i)z}{}^c(z)\mathscr{D}_{(i)} + \frac{1}{i}\sum_i \lambda_{z(i)}{}^a(z)\mathscr{D}_{(i)}\frac{\delta}{\delta A_{\bar{z}}{}^a(z)}\right). \tag{2.4.1}$$

Now we use (2.33) again. The resulting expression can be simplified by setting F to zero, and using the fact that $\mathscr{D}_{(i)}F = 0$ at $F = 0$ (by virtue of (2.39) and (2.27)). Also it is convenient to integrate by parts in the w variable in the first line of (2.41), using the delta function to eliminate the

w integration. After these steps, the connection on $\tilde{\mathscr{H}}_Q$ turns out to be

$$\delta^{\tilde{\mathscr{H}}_Q} = \delta - t\mathscr{O}, \tag{2.42.1}$$

$$\begin{aligned} \mathscr{O} = -\frac{\pi}{k}\int_\Sigma \delta J_{\bar{z}}{}^{z} \Bigg(& \frac{k}{4\pi}(\delta^b{}_a D_w L_z{}^a{}_b(z, w))|_{w=z} \\ & - L_z{}^a{}_d(z, z) f_{ac}{}^d \lambda_{(i)z}{}^c(z)\mathscr{D}_{(i)} \\ & + \frac{1}{i}\sum_i \lambda_{(i)z}{}^a(z)\mathscr{D}_{(i)} \sum_j \lambda_{(j)z}{}^a(z)\mathscr{D}_{(j)} \Bigg). \end{aligned} \tag{2.42.2}$$

Notice that the combination

$$\sum_i \lambda_{(i)}{}^a(z)\mathscr{D}_{(i)}, \tag{2.43}$$

which is the only expression through which the $\lambda_{(i)}$ appear in (2.42), is independent of the choice of an orthonormal basis of the $\lambda_{(i)}{}^a$.

Regularization. The first problem in understanding (2.42) is to make sense of the Green's functions on the diagonal

$$f_{ac}{}^d L_z{}^a{}_d(z, z) \quad \text{and} \quad \left(\delta^b{}_a D_w L_z{}^a{}_b(z, w)\right)|_{w=z}. \tag{2.44}$$

These particular Green's functions on the diagonal have an interpretation familiar to physicists. Consider a free field theory with an anticommuting spin zero field c and a $(1, 0)$-form b, both in the adjoint representation, and with the Lagrangian

$$\mathscr{L} = \int_\Sigma d\bar{z}dz D_{\bar{z}}c^a(z)b_z^a(z). \tag{2.45}$$

Let us introduce the current $J_{zc}(z) = f_{ca}{}^d b_z^a(z)c^d(z)$ and the stress tensor $T_{zz}(z) = b_z^a(z)D_z c^a(z)$. Then the Green's functions appearing in (2.44) are formally

$$f_{ca}{}^d L_z{}^a{}_d(z, z) = \langle J_{zc}\rangle', \tag{2.46}$$

and

$$\left(\delta^b{}_a D_w L_z{}^a{}_b(z, w)\right)|_{w=z} = \langle T_{zz}(z)\rangle', \tag{2.47}$$

where the symbol $\langle\ \rangle'$ means to take an expectation value with the kernel of the kinetic operator D_z projected out. Interpreted in this way, the desired Green's functions on the diagonal have been extensively studied in the physics literature on "anomalies". Thus, the crucial properties of these particular Green's functions are well known to physicists.

The interpretation of the Green's functions on the diagonal that appear in (2.44) as the expectation values of the current J^a and the stress tensor T can also be recast in a language that will be recognizable to mathematicians. Introduce a metric $g_{\mu\nu}$ on Σ which is a Kähler metric for the complex structure J. Then one has the Laplacian $\Delta : \Gamma(\Sigma, \mathrm{ad}(E)) \to \Gamma(\Sigma, \mathrm{ad}(E))$, defined by $\Delta = -g^{\mu\nu} D_\mu D_\nu$. Let

$$H = \det{}'(\Delta) \tag{2.48}$$

be the regularized determinant of the Laplacian. H is a functional of the connection A (which appears in the covariant derivative D_μ) and the metric g. We can interpret (2.46) and (2.47) as the statements that

$$f_{ca}{}^{d} L_z{}^{a}{}_{d}(z, z) = -i \frac{\delta}{\delta A_{\bar{z}}{}^{c}(z)} \ln H, \tag{2.49}$$

and

$$(\delta^b{}_a D_w L_z{}^a{}_b(z, w))|_{w=z} = \frac{\delta}{\delta g^{zz}} \ln H. \tag{2.50}$$

Given a regularization of the determinant of the Laplacian—for instance, Pauli-Villars regularization, often used by physicists, or zeta function regularization, usually preferred in the mathematical theory—the right-hand sides of (2.49) and (2.50) are perfectly well defined, and can serve as definitions of the left-hand sides. Since these were the problematic terms in the formula (2.42) for the connection on the quantum bundle, we have now made this formula well defined. It remains to determine whether this connection has the desired properties.

Conformal and diffeomorphism invariance. In defining the Green's functions on the diagonal, we have had to introduce a metric, not just a complex structure. To ensure diffeomorphism invariance we will choose the metric to depend on the complex structure in a natural way. Before making such a choice, however, we shall explain the simple way in which the connection (2.42) transforms under conformal rescalings of the metric.

One knows from the theory of regularized determinants (or the theory of the conformal anomaly in $(1+1)$-dimensions) that under a conformal rescaling $g \to e^{\phi} g$ of the metric, with ϕ being a real-valued function on Σ, the regularized determinant H transforms as

$$H \to \exp(S(\phi, g)) \cdot H, \tag{2.51}$$

where $S(\phi, g)$ is the Liouville action with an appropriate normalization and may be defined by

$$\frac{\partial}{\partial\phi}S(0, g) = \frac{\mathrm{Dim}(G)}{24\pi}\sqrt{g}R, \tag{2.52}$$

together with the group laws $S(0, g) = 1$ and $S(\phi_1+\phi_2, g) = S(\phi_1, e^{\phi_2}g)\cdot S(\phi_2, g)$. Here, $\sqrt{g}R$ is the scalar curvature density of the metric g. The crucial property of $S(\phi, g)$ is that it is independent of the connection A. We conclude that under a conformal rescaling of the metric, the current expectation value defined in (2.49) is invariant. On the other hand, the expectation value (2.50) of the stress tensor transforms as

$$\begin{aligned}\left(\delta^b{}_a D_w L_z{}^a{}_b(z, w)\right)|_{w=z} \to \left(\delta^b{}_a D_w L_z{}^a{}_b(z, w)\right)|_{w=z} \\ + \frac{\delta}{\delta g^{zz}}(S(\phi, g)).\end{aligned} \tag{2.53}$$

This means that under a conformal transformation of the metric of Σ, the connection $\delta^{\tilde{\mathscr{H}}_Q}$ defined in (2.42) transforms as

$$\delta^{\tilde{\mathscr{H}}_Q} \to \delta^{\tilde{\mathscr{H}}_Q} + t\pi\int_\Sigma \delta J_{\bar{z}}{}^z \frac{\delta}{\delta g^{zz}}(S(\phi, g)). \tag{2.54}$$

The second term on the right-hand side of (2.54), being independent of the connection A, is a function on the base in the fibration $\mathscr{M}\times\mathscr{T}\to\mathscr{T}$. On each fiber, this function is a constant, and this means that the second term on the right in (2.54) is a central term. Up to a projective factor, $\delta^{\tilde{\mathscr{H}}_Q}$ is conformally invariant. The central term in the change of $\delta^{\tilde{\mathscr{H}}_Q}$ under a conformal transformation has for its $(1+1)$-dimensional counterpart the conformal anomaly in current algebra.

To show that our connection lives over Teichmüller space and is invariant under the mapping class group one must check that $\delta^{\tilde{\mathscr{H}}_Q}$ is invariant under a diffeomorphism of Σ, and that the connection form $\mathscr{O}$ vanishes for the variation $\delta J_{\bar{z}}{}^z = \partial_{\bar{z}}v^z$ of the complex structure induced by a vector field v on Σ. The first assertion is automatic if we always equip Σ with a metric that is determined in a natural way by the complex structure (for instance, the constant curvature metric of unit area or the Arakelov metric), since except for the choice of metric the rest of our construction is natural and so diffeomorphism invariant. The second point may be verified directly by substituting $\delta J_{\bar{z}}{}^z = \partial_{\bar{z}}v^z$ into (2.42), integrating by parts, and using (2.28). It follows more conceptually from the fact that the connection may be written in a form intrinsic on $\mathscr{M}$ (see §4) and from the fact that H and the Kähler structure on $\mathscr{M}$ depend on the complex structure J on Σ only up to isotopy.

Properties to be verified. Let us state precisely what has been achieved so far. Over the moduli space $\mathscr{M}$ of flat connections, we fix a Hermitian line bundle $\tilde{\mathscr{L}}$ of curvature $-i\tilde{\omega}$ with an action of the mapping class group. The prequantum Hilbert space is $\Gamma(\mathscr{M}, \tilde{\mathscr{L}})$. The prequantum bundle over Teichmuller space $\mathscr{T}$ is the trivial bundle $\tilde{\mathscr{H}}_{\mathrm{pr}} = \Gamma(\mathscr{M}, \mathscr{L}) \times \mathscr{T}$. The connection $\delta - t\mathscr{O}$ (with the regularization defined in (2.49), (2.50)) is rigorously well defined as a connection on the prequantum bundle. What remains is to show that this connection has the following desired properties.

(i) The quantum bundle $\tilde{\mathscr{H}}_Q$ over $\mathscr{T}$ is the subbundle of $\tilde{\mathscr{H}}_{\mathrm{pr}}$ consisting of *holomorphic* sections of the prequantum line bundle; that is, $\tilde{\mathscr{H}}_Q|_{J_t} = H^0(\mathscr{M}_J, \tilde{\mathscr{L}})$. We would like to show that (with the correct choice of the parameter t) the connection $\delta - t\mathscr{O}$ on $\tilde{\mathscr{H}}_{\mathrm{pr}}$ preserves holomorphicity and thus restricts to a connection on $\tilde{\mathscr{H}}_Q$.

(ii) We would like to show that this restricted connection, $\delta^{\tilde{\mathscr{H}}_Q}$, is projectively flat, and that it is unitary for the correct choice of a unitary structure on $\tilde{\mathscr{H}}_Q$.

Of course, for symplectic quotients of finite-dimensional affine spaces, these properties would be automatic consequences of simple "upstairs" facts that are easy to verify. The reason that there is something to be done is that here the underlying affine space is infinite dimensional. Since we do not have a rigorous quantization of the "upstairs" space, we need to verify *ex post facto* that the connection $\delta^{\tilde{\mathscr{H}}_Q}$ has the desired properties. Except for unitarity, which we will not be able to understand except in genus one (see §5), this will be done in §§3, 4, and 7. The computations will be done in a framework that is expressed directly, to the extent possible, in terms of the intrinsic geometry of the moduli space $\mathscr{M}$. These computations could be carried out directly in the framework and notation of the present section, but they are simpler if expressed in terms of the intrinsic geometry of $\mathscr{M}$. However, we will here describe (nonrigorously, but in a language that may be quite familiar to some readers) a small piece of the direct, explicit verification of property (i). This piece of the verification of (i) is illuminating because it explains a phenomenon that is well known in conformal field theory, namely the replacement of the "level" k by $k+h$ in many formulas.

The anomaly. In (2.30) and (2.31), we have introduced bases $\mathscr{D}_{(i)}$ and $\overline{\mathscr{D}}_{(m)}$ of $T^{(1,0)}\mathscr{M}$ and $T^{(0,1)}\mathscr{M}$, respectively. The symplectic structure of $\mathscr{M}$ can be described by the statement that, acting on sections of $\tilde{\mathscr{L}}$,

$$[\overline{\mathscr{D}}_{(m)}, \mathscr{D}_{(i)}] = -\frac{k}{4\pi}\delta_{im}. \tag{2.55}$$

This can be verified directly using (2.18) and the orthonormality of the λ's (and the fact that a section of $\tilde{\mathscr{L}}$ over $\mathscr{M}$ is the same as a $\mathscr{G}_c$-invariant section on $\mathscr{A}$).

Now, property (i) above—that the connection $\delta - t\mathscr{O}$ preserves holomorphicity—amounts to the statement that, at least when acting on holomorphic sections of $\tilde{\mathscr{L}}$,

$$0 = [\delta - t\mathscr{O}, \overline{\mathscr{D}}_{(m)}] = [\delta, \overline{\mathscr{D}}_{(m)}] + t[\overline{\mathscr{D}}_{(m)}, \mathscr{O}]. \tag{2.56}$$

The analogue of (2.56) would of course be true for symplectic quotients of finite-dimensional affine spaces. For the present problem, (2.56) can be verified directly although tediously. In doing so, one meets many terms that would be present in the finite-dimensional case. There is really only one point at which one meets an "anomaly" that would not be present in the finite-dimensional case. This comes from the term in $t[\overline{\mathscr{D}}_{(m)}, \mathscr{O}]$ with the structure

$$-\frac{t\pi}{k}\int_\Sigma \delta J_{\bar z}{}^{z}\left(\overline{\mathscr{D}}_{(m)}\langle J_{za}(z)\rangle'\right)\sum_{(i)}\lambda_{(i)z}{}^{a}(z)\mathscr{D}_{(i)}. \tag{2.57}$$

Now, formally the current $J_z{}^a(z)$ is defined in a Lagrangian (2.45) that depends holomorphically on the connection A, that is, $A_{\bar z}$ and not A_z appears in this Lagrangian. Naively, therefore, one might expect that $< J_{za}(z) >'$ or any other quantity computed from this Lagrangian would be independent of A_z and therefore annihilated by $\overline{\mathscr{D}}_{(m)}$.

However, the quantum field theory defined by the Lagrangian (2.45) is anomalous. As a result of the anomaly in this theory, there is a clash between gauge invariance and the claim that the current is independent of A_z. At least for Σ of genus zero, where there are no zero modes to worry about, one can indeed define the quantum current $< J_z{}^a(z) >'$ so as to be annihilated by $\delta/\delta A_z{}^a(z)$, but in this case $< J_{za} >'$ is not gauge invariant. In the case at hand, we must insist on gauge invariance since otherwise the basic formulas such as the definition of the connection (2.42) do not make sense on the moduli space $\mathscr{M}$. Indeed, in (2.49) we have regulated the current in a way that preserves gauge invariance. The anomaly is the assertion that the gauge invariant current defined in (2.49) cannot be independent of A_z; rather one has

$$\frac{\delta}{\delta A_w{}^b(w)} < J_{za}(z) >' = \frac{1}{2\pi}\delta_{ab}\delta_{z\overline{w}}(z, w)h + \cdots . \tag{2.58}$$

Here h is the dual Coxeter number, defined in (2.3). The $\cdots$ terms in (2.58), which would be absent for Σ of genus zero, arise because in addition to the anomalous term that comes from the short distance anomaly in quantizing the chiral Lagrangian (2.45), there is an additional dependence of $\langle J_z\rangle'$ on A_z that comes because of projecting away the zero modes present in (2.45) in defining $\langle\ \rangle'$. These $\cdots$ terms have analogs for symplectic quotients of finite-dimensional manifolds, and cancel in a somewhat elaborate way against other terms that arise in evaluating (2.56). We want to focus on the implications of the anomalous term.

The contribution of the anomalous term to (2.57) and (2.56) is

$$-\left(\frac{ith}{2k}\right)\int_\Sigma \delta J_{\overline{z}}{}^z \lambda_{(m)z}{}^a(z)\sum_{(i)}\lambda_{(i)z}{}^a(z)\mathscr{D}_{(i)}. \tag{2.59}$$

Terms of the same structure come from two other sources. As we see in (2.42), the last term in the connection form $\mathscr{O}$ is a second-order differential operator $\mathscr{O}_2$. In computing $t[\overline{\mathscr{D}}_{(m)}, \mathscr{O}_2]$, one finds with the use of 2.55 a term

$$-\left(\frac{it}{2}\right)\int_\Sigma \delta J_{\overline{z}}{}^z \lambda_{(m)z}{}^a(z)\sum_{(i)}\lambda_{z(i)}{}^a(z)\mathscr{D}_{(i)}. \tag{2.60}$$

The last contribution of a similar nature comes from

$$\begin{aligned}[\delta, \overline{\mathscr{D}}_{(m)}] &= \left[\delta, \int_\Sigma \lambda_{(m)z}{}^a(z)\frac{\delta}{\delta A_z{}^a(z)}\right] \\ &= \int_\Sigma \lambda_{(m)z}{}^a(z)\left[\delta, \frac{\delta}{\delta A_z{}^a(z)}\right] + \cdots .\end{aligned} \tag{2.61}$$

The $\cdots$ terms are proportional to $\delta/\delta A_z$ and annihilate holomorphic sections of $\mathscr{L}$. On the other hand,

$$\left[\delta, \frac{\delta}{\delta A_z{}^a(z)}\right] = \frac{i}{2}\delta J_{\overline{z}}{}^z \frac{\delta}{\delta A_{\overline{z}}{}^a(z)}. \tag{2.62}$$

Therefore, on holomorphic sections, after using (2.33), we get

$$[\delta, \overline{\mathscr{D}}_{(m)}] = \frac{i}{2}\int_\Sigma \delta J_{\overline{z}}{}^z \lambda_{(m)z}{}^a(z)\sum_i \lambda_{(i)z}{}^a(z)\mathscr{D}_{(i)}. \tag{2.63}$$

In the absence of the anomalous term (2.59), the two terms (2.63) and (2.60) would cancel precisely if $t = 1$. This is why the correct value in the quantization of a finite-dimensional affine space is $t = 1$. However, including the anomalous term, (2.59), the three expressions (2.59), (2.60), and (2.63) sum to zero if and only if $t = k/(k+h)$. The connection on the quantum bundle $\mathscr{H}_Q \to \mathscr{T}$ is thus finally pinned down to be

$$\begin{aligned}\nabla = \delta + \frac{\pi}{k+h}\int_\Sigma \Bigg(& \frac{k}{4\pi}(\delta^b{}_a D_w L_z{}^a{}_b(z, w))|_{w=z} \\ & - L_z{}^a{}_d(z, z) f_{ac}{}^d \lambda_{(i)z}{}^c(z)\mathscr{D}_{(i)} \\ & + \frac{1}{i}\sum_i \lambda_{(i)z}{}^a(z)\mathscr{D}_{(i)} \sum_j \lambda_{(j)z}{}^a(z)\mathscr{D}_{(j)} \Bigg).\end{aligned} \tag{2.64}$$

That the connection form is proportional to $1/(k+h)$ rather than $1/k$, which one would obtain in quantizing a finite dimensional affine space, is analogous to (and can be considered to explain) similar phenomena in two-dimensional conformal field theory.

The rest of the verification of (2.56) is tedious but straightforward. No further anomalies arise; the computation proceeds just as it would in the quantization of a finite-dimensional affine space. We forego the details here, since we will give a succinct and rigorous proof of (2.56) in §4.

Commun. Math. Phys. 144, 189–212 (1992)

Communications in
Mathematical
Physics

On Holomorphic Factorization of WZW and Coset Models

Edward Witten*

School of Natural Sciences, Institute for Advanced Study, Olden Lane, Princeton, NJ 08540, USA

Received July 19, 1991

Abstract. It is shown how coupling to gauge fields can be used to explain the basic facts concerning holomorphic factorization of the WZW model of two dimensional conformal field theory, which previously have been understood primarily by using conformal field theory Ward identities. We also consider in a similar vein the holomorphic factorization of G/H coset models. We discuss the G/G model as a topological field theory and comment on a conjecture by Spiegelglas.

1. Introduction

The WZW model of two dimensional conformal field theory [1] is a quantum field theory in which the basic field is a map $g:\Sigma\to G$, Σ being a two dimensional Riemann surface and G being a compact Lie group, which for simplicity we will in this paper take to be simple, connected and simply connected. The basic WZW functional is

$$I(g)=-\frac{1}{8\pi}\int_{\Sigma} d^2\sigma\sqrt{\varrho}\varrho^{ij}\operatorname{Tr}(g^{-1}\partial_i g\cdot g^{-1}\partial_j g)-i\Gamma(g), \tag{1.1}$$

where ϱ is a metric on Σ, Tr is an invariant form on the Lie algebra of G whose normalization will be specified presently, and Γ is the Wess-Zumino term [2]. The latter has the following description [3] in case Σ is a Riemann surface without boundary. (For the more general case see [4].) Let B be a three manifold such that $\partial B=\Sigma$, pick an extension of g over B, which we will also call g, and let

$$\Gamma(g)=\int_B g^*\omega=\frac{1}{12\pi}\int_B d^3\sigma\varepsilon^{ijk}\operatorname{Tr} g^{-1}\partial_i g\cdot g^{-1}\partial_j g\cdot g^{-1}\partial_k g, \tag{1.2}$$

where ω is the left and right invariant three form on the G manifold defined by

$$\omega=\frac{1}{12\pi}\operatorname{Tr}(g^{-1}dg\wedge g^{-1}dg\wedge g^{-1}dg). \tag{1.3}$$

---

* Research supported in part by NSF Grant PHY86-20266

$\Gamma(g)$ is well defined (independent of the choice of B and the extension of g over B) modulo the periods of ω. In these formulas, "Tr" is an invariant quadratic form on the Lie algebra of G which we take to be the smallest multiple of the trace in the adjoint representation such that the periods of ω are multiples of 2π. (For $G=SU(N)$, "Tr" is simply the trace in the N dimensional representation.) The condition on the periods ensures that the WZW functional $I(g)$ is well-defined as a map to $\mathbb{C}/2\pi i\mathbb{Z}$, so that $e^{-I(g)}$ is well-defined as a complex valued functional on the space of maps $\Sigma\to G$.

The Lagrangian of the WZW model is $L(g)=kI(g)$, where k is a positive integer called the "level," and the partition function Z of the WZW model is formally defined as a Feynman path integral,

$$Z=\int Dg\, e^{-L}=\int Dg\, e^{-kI}\,. \tag{1.4}$$

Z depends, of course, on the metric ϱ of Σ which enters in the definition of I. Conformal invariance of the WZW model means that apart from a relatively elementary factor given by the conformal anomaly, Z depends only on the complex structure determined by ϱ.

The WZW model is essentially exactly soluble; the ability to solve it depends on its holomorphic factorization, first investigated by Knizhnik and Zamolodchikov [5]. Holomorphic factorization of the WZW model means that locally on the space of complex structures one can write $Z=\sum_i f_i\bar{f}_i$, where the f_i are holomorphic functions on the space of complex structures. Globally, as advocated by Friedan and Shenker [6], one interprets this formula to mean $Z=(f,f)$, where f is a holomorphic section of a certain flat vector bundle $\mathscr{V}$ over moduli space equipped with a hermitian form (,). (The conformal anomaly means that these statements require a somewhat more precise formulation.) The flat bundles that arise in this way have been extensively studied [7, 8] and have been seen to have a natural origin in gauge theories [9–12]. To date, the existence of a holomorphic factorization of the WZW model has mostly been understood using conformal field theory Ward identities, this being the original point of view of Knizhnik and Zamolodchikov. The purpose of the present paper is to use gauge theories – or more exactly, coupling of the WZW model to gauge fields – to deduce the existence of a holomorphic factorization. Many of the key steps have been previously exploited by Gawedzki and Kupianen [13, 14]. See also the work of Bernard [15] on the heat equation obeyed by characters of affine Lie algebras. The main novelty which motivated me to write the present paper is the integration over the gauge field and use of the Polyakov-Wiegmann formula to prove that the partition function has a holomorphic factorization; see the derivation of Eq. (2.28). Our treatment will be formal; we will make no claim to analyze the quantum anomalies.

Gauged WZW models have been extensively studied [16–18, 14], mainly with the aim of giving a Lagrangian description of the GKO coset models [19] (whose prehistory goes back to the early days of string theory [20]). After developing our approach to holomorphic factorization of the WZW model in Sect. 2, we will apply the same methods to holomorphic factorization of coset models in Sect. 3 (recovering observations of Moore and Seiberg [21] and Gawedzki and Kupianen [13, 14]), and then we will consider the special case of the G/G coset model, where sharper statements can be made, as this theory is actually a topological field theory. The G/G model has been investigated by Spiegelglas [22].

2. Gauge Couplings and Holomorphic Factorization

The WZW action functional $I(g)$ is invariant under the usual action of $G\times G$ (often called $G_L\times G_R$) on the G manifold. An element (a,b) of $G_L\times G_R$ acts on G by $g\to agb^{-1}$.[1] Given a Lagrangian with a (global) symmetry, it is usually possible to "gauge" the symmetry, introducing gauge fields and constructing a gauge invariant extension of the original Lagrangian. In particular, gauging the WZW model means generalizing the theory from the case in which g is a map $\Sigma\to G$ to the case in which g is a section of a bundle $X\to\Sigma$ with fiber G and structure group $G_L\times G_R$ or a subgroup. Letting A be a connection on such a bundle, one aims to find a gauge invariant functional $I(g,A)$ (whose variation with respect to g or A is required to be given by a local formula) which reduces, for X the trivial bundle and $A=0$, to $I(g)$.

In the case of the WZW model, such a gauge invariant extension does not exist. There is no problem in gauging the first term in (1.1) – one just replaces derivatives by covariant derivatives. However, the Wess-Zumino term $\Gamma(g)$ does not have a gauge invariant extension unless one restricts to an "anomaly-free" subgroup F of $G_L\times G_R$ (and considers bundles $X\to\Sigma$ with structure group F). The condition for a subgroup to be anomaly free can bc stated as follows. For any subgroup F of $G_L\times G_R$, $\mathscr{G}_L$ and $\mathscr{G}_R$ (the adjoint representations of G_L and G_R) are F modules. If Tr_L and Tr_R are the traces in $\mathscr{G}_L$ and $\mathscr{G}_R$, then the condition for absence of anomalies is that for any $t,t'\in\mathscr{F}$ (the Lie algebra of F)

$$\mathrm{Tr}_L tt'-\mathrm{Tr}_R tt'=0\,. \tag{2.1}$$

(As will be clear in the appendix, this statement is equivalent to the statement that the class in $H^3(G,\mathbb{R})$ represented by ω has an extension in $H^3_F(G,\mathbb{R})$, where H^*_F is the F-equivariant cohomology.) In the appendix, we will review the derivation of (2.1) for the benefit of readers not already familiar with such matters and to clarify a few geometrical points. Some readers may want to consult the appendix first, but this should not be necessary for readers who are familiar with gauged WZW models or are willing to verify by hand a few easily verified formulas.

We will also be interested in gauged WZW actions in cases in which (2.1) is *not* obeyed. In such a case, one cannot construct a gauge invariant $I(g,A)$, but one can find a "best possible" $I(g,A)$, such that the violation of gauge invariance is a multiple of a standard "anomaly" expression that depends on A but not on g. [A topological explanation of why it is possible to do this will be given in the appendix, where the detailed formula for $I(g,A)$ is also explained.]

We will adopt the following conventions: z will be a local complex coordinate on Σ, d^2z is the measure $|dzd\bar z|$, and the components of A are defined by $A=A_z dz+A_{\bar z}d\bar z$. We sometimes use the Levi-Civita antisymmetric tensor density ε^{ij} with $\varepsilon^{z\bar z}=-\varepsilon^{\bar z z}=i$. (That is, for one forms a and b, we write $\int_\Sigma a\wedge b=\int_\Sigma d^2z\varepsilon^{ij}a_ib_j$.) Our orientation conventions can be most efficiently and usefully explained by saying that the variation of the Wess-Zumino term under $\delta g=-gu$ is

$$\delta\Gamma=-\frac{1}{4\pi}\int_\Sigma d^2z\varepsilon^{ij}\,\mathrm{Tr}(ug^{-1}\partial_i g\cdot g^{-1}\partial_j g)\,. \tag{2.2}$$

[1] If G has a non-trivial center $Z(G)$, then $Z(G)$, diagonally embedded in $G_L\times G_R$, acts trivially on g, so the faithfully acting symmetry group is really $(G_L\times G_R)/Z(G)$. This refinement will not be important until we come to coset models

For simplicity, we will in this paper consider only the case that G is connected and simply connected, so that a G bundle over Σ is trivial.

2.1. The Holomorphic Wave-Function

To begin with, we consider the case that $F = G_R$. In this case, (2.1) is not obeyed, so a gauge invariant functional $I(g, A)$ does not exist. However, we take

$$I(g, A) = I(g) + \frac{1}{2\pi} \int_\Sigma d^2z \, \mathrm{Tr}\, A_{\bar z} g^{-1} \partial_z g - \frac{1}{4\pi} \int_\Sigma d^2z \, \mathrm{Tr}\, A_{\bar z} A_z , \tag{2.3}$$

which is as close as there is to a gauge invariant functional, in the following sense. Under an infinitesimal gauge transformation,

$$\delta g = -gu , \qquad \delta A_i = -D_i u = -\partial_i u - [A_i, u] , \tag{2.4}$$

one has

$$\delta I(g, A) = \frac{1}{4\pi} \int_\Sigma d^2z \, \mathrm{Tr}\, u(\partial_z A_{\bar z} - \partial_{\bar z} A_z) = \frac{i}{4\pi} \int_\Sigma \mathrm{Tr}\, u \, dA , \tag{2.5}$$

an expression which depends on A but not on g or the complex structure of Σ. Equation (2.3) is the unique extension of $I(g)$ with this property.

We now formally define a functional of A by

$$\begin{aligned} \Psi(A) &= \int Dg \, e^{-kI(g, A)} \\ &= \int Dg \exp\left(-kI(g) - \frac{k}{2\pi} \int_\Sigma d^2z \, \mathrm{Tr}\, A_{\bar z} g^{-1} \partial_z g + \frac{k}{4\pi} \int_\Sigma d^2z \, \mathrm{Tr}\, A_{\bar z} A_z \right). \end{aligned} \tag{2.6}$$

Note that we do not treat A as a quantum variable; that is, we do not integrate over A. This would not be sensible as $I(g, A)$ is not gauge invariant.

Now, Ψ obeys two key equations. First,

$$\left(\frac{\delta}{\delta A_z} - \frac{k}{4\pi} A_{\bar z} \right) \Psi = 0 , \tag{2.7}$$

and second

$$\left(D_{\bar z} \frac{\delta}{\delta A_{\bar z}} + \frac{k}{4\pi} D_z A_{\bar z} - \frac{k}{2\pi} F_{\bar z z} \right) \Psi = 0 . \tag{2.8}$$

Equation (2.7) is proved simply by acting with the left-hand side on the integral representation of Ψ, and differentiating under the integral sign. Equation (2.8) is a consequence of the standard anomaly Eq. (2.5). By differentiating under the integral sign, one finds that the left-hand side of (2.8) equals

$$-\frac{k}{2\pi} \int Dg \, e^{-kI(g, A)} [D_{\bar z}(g^{-1} D_z g) + F_{\bar z z}] , \tag{2.9}$$

where we have introduced covariant derivatives $Dg = dg - gA$. The quantity in brackets in (2.9) is the equation of motion of the g field – the variation of $I(g, A)$ under $\delta g = -gu$. Therefore, upon integrating over g, (2.9) vanishes, by integration by parts in g space.

To elucidate these equations, it is useful to first introduce the operators

$$\begin{aligned}\frac{D}{DA_z}&=\frac{\delta}{\delta A_z}-\frac{k}{4\pi}A_{\bar z}\\ \frac{D}{DA_{\bar z}}&=\frac{\delta}{\delta A_{\bar z}}+\frac{k}{4\pi}A_z.\end{aligned}\tag{2.10}$$

Note that for $z, w\in\Sigma$,

$$\left[\frac{D}{DA_z(z)},\frac{D}{DA_{\bar w}(w)}\right]=\frac{k}{2\pi}\delta(z,w).\tag{2.11}$$

In terms of these operators, (2.7) is simply

$$\frac{D}{DA_z}\Psi=0,\tag{2.12}$$

and (2.8) becomes

$$\left(D_{\bar z}\frac{D}{DA_{\bar z}}-\frac{k}{2\pi}F_{\bar z z}\right)\Psi=0,\tag{2.13}$$

which in view of (2.12) can be written in a way that does not refer to the complex structure of Σ:

$$\left(D_i\frac{D}{DA_i}-\frac{ik}{4\pi}\varepsilon^{ij}F_{ij}\right)\Psi=0.\tag{2.14}$$

These equations are closely related to the basic formulas that appear in canonical quantization of 2+1 dimensional Chern-Simons gauge theory, as explained, for instance, in Sect. 2 of [11] or in [10]. Let $\mathscr{A}$ be the space of connections on (the trivial F bundle over) Σ. $\mathscr{A}$ has a symplectic structure that can be defined purely in differential topology, without choosing a conformal structure on Σ. It can be defined by the formula

$$\omega(a_1,a_2)=\frac{1}{2\pi}\int_\Sigma \operatorname{Tr} a_1\wedge a_2,\tag{2.15}$$

where a_1 and a_2 are any two adjoint-valued one forms representing tangent vectors to $\mathscr{A}$. "Prequantization" of $\mathscr{A}$ (in the sense of Kostant [23] and Souriau [24]) means constructing a unitary complex line bundle $\mathscr{L}$ with a connection whose curvature is $-i\omega$. Equation (2.10) can be regarded as a formula defining a connection on the trivial complex line bundle $\mathscr{P}=\mathscr{A}\times\mathbb{C}$ over $\mathscr{A}$ (which we take with the standard unitary structure). This connection according to (2.11) has curvature $-ik\omega$. The factor of k means that $\mathscr{P}$, with this connection, can be identified as $\mathscr{L}^{\otimes k}$, with $\mathscr{L}$ the basic prequantum line bundle.

Hence (2.10) and (2.11) actually describe prequantization of $\mathscr{A}$, with the symplectic structure $k\omega$. The notion of prequantization obviously does not depend on a choice of polarization or complex structure, and indeed, though (2.10) and (2.11) are written in terms of a local complex coordinate on Σ, they are actually entirely independent of the conformal structure of Σ. $\Psi(A)$ is best regarded not as a "function" on $\mathscr{A}$ but as a section of the prequantum line bundle $\mathscr{L}^{\otimes k}$.

The complex structure enters when one wishes to *quantize* $\mathscr{A}$. A choice of complex structure on Σ determines a complex structure on $\mathscr{A}$ – in which the $A_{\bar z}$ are holomorphic and the A_z are antiholomorphic. This complex structure is a "polarization" which permits quantization: the quantum Hilbert space consists of *holomorphic* sections of $\mathscr{L}^{\otimes k}$. Equation (2.12) means simply that $\Psi(A)$ is such a holomorphic section.

Now let us discuss the meaning of (2.13). Let $\hat{F}$ be the group of gauge transformations, that is, the group of maps of Σ to F. Acting on *functions* on $\mathscr{A}$, $\hat{F}$ is generated by the operators

$$D_i \frac{\delta}{\delta A_i}. \tag{2.16}$$

To find a $\hat{F}$ action on sections of $\mathscr{L}^{\otimes k}$, one must "lift" the vector fields in (2.16) appropriately. This can be done in a standard fashion (for instance, see Sect. 2 of [11]); the $\hat{F}$ action on sections of $\mathscr{L}$ is generated by the operators

$$D_i \frac{D}{DA_i} - \frac{ik}{4\pi} \varepsilon^{ij} F_{ij}. \tag{2.17}$$

(The second term is the contribution of the classical "moment map.") We can thus see the meaning of (2.13) or (2.14) – Ψ is gauge invariant, as a section of $\mathscr{L}^{\otimes k}$.

The two conditions we have found – that Ψ is holomorphic and gauge invariant – mean together that Ψ can be regarded as a physical state of $2+1$ dimensional Chern-Simons gauge theory (with gauge group F). (See [9–11] for more background.) This in fact can be regarded as the essential relation between the WZW model and Chern-Simons theory. We will now recall a few further facts about the Chern-Simons theory. (The facts summarized in the next three paragraphs are not all strictly needed for reading the present paper, but they help put the discussion in context.)

The $\hat{F}$ action on sections of $\mathscr{L}^{\otimes k}$ does not depend on the conformal structure of Σ, but something new happens once such a conformal structure is picked. A connection A on (the trivial F bundle over) a complex Riemann surface Σ determines an operator $\bar{\partial}_A$ which defines a complex structure on the bundle. Gauge transformations act by $\bar{\partial}_A \to f \bar{\partial}_A f^{-1}$, for $f: \Sigma \to F$, but as this formula makes sense for $f: \Sigma \to F_{\mathbb{C}}$ ($F_{\mathbb{C}}$ is the complexification of F), one actually gets an action of $\hat{F}_{\mathbb{C}}$ (the group of maps of Σ to $F_{\mathbb{C}}$) on $\mathscr{A}$. A $\hat{F}$ invariant section of $\mathscr{L}^{\otimes k}$ which is also holomorphic is automatically $\hat{F}_{\mathbb{C}}$ invariant. Let V be the space of holomorphic, $\hat{F}_{\mathbb{C}}$ invariant sections of $\mathscr{L}^{\otimes k}$. V is the space of physical states in Chern-Simons gauge theory, at level k. From what we have said above, Ψ is a vector in V.

A $\hat{F}$ invariant section of $\mathscr{L}^{\otimes k}$ is the same as a section of an appropriate push-down line bundle, which we will also call $\mathscr{L}^{\otimes k}$, over the quotient space $\mathscr{A}/\hat{F}_{\mathbb{C}}$. The quotient $\mathscr{A}/\hat{F}_{\mathbb{C}}$, with the quotient taken in an appropriate sense, is the moduli space of stable holomorphic $F_{\mathbb{C}}$ bundles over Σ, or (by a theorem of Narasimhan and Seshadri) the moduli space $\mathscr{M}$ of flat F connections on Σ, up to gauge transformation. This is a compact complex manifold, and in particular, the vector space V, which can be identified as $H^0(\mathscr{M}, \mathscr{L}^{\otimes k})$, is finite dimensional.

So far, when we have made statements that depend on the complex structure of Σ, we have considered Σ with a *fixed* complex structure. Permitting the complex structure of Σ to vary, we get not a single vector space V but a family of vector spaces parametrized by the space $\mathscr{S}$ of complex structures on Σ, or in short, a vector bundle $\mathscr{V}$ over $\mathscr{S}$. The bundle $\mathscr{V} \to \mathscr{S}$ has a natural projectively flat connection (which is essential for the topological invariance of Chern-Simons theory); the holomorphic structure is obvious, and the anti-holomorphic structure, which we will recall at an appropriate point, is less obvious.

2.2. The Norm of the Wave Function

By now we have defined, for every complex Riemann surface Σ, a vector space V consisting of holomorphic, gauge invariant sections of the line bundle $\mathscr{L}^{\otimes k}$ over the space $\mathscr{A}$ of connections. A natural Hermitian structure on V is given formally by

$$(\Psi_1, \Psi_2) = \frac{1}{\mathrm{vol}(\hat{G})} \int_{\mathscr{A}} DA\, \overline{\Psi_1(A)}\, \Psi_2(A). \tag{2.18}$$

(Formally, DA is the measure on $\mathscr{A}$ determined by its symplectic structure, and it is natural to divide by the volume of $\hat{F} \cong \hat{G}$ because of the gauge invariance of $\bar{\Psi}_1 \Psi_2$.) In genus one, this Hermitian structure can be worked out explicitly (that is, reduced to an explicit description of an inner product on the finite dimensional vector space V), by actually computing the integral over the infinite dimensional $F_{\mathbb{C}}$ orbits [14, 25]. In genus >1, such an explicit evaluation is not known.

We want to compute the norm of the vector Ψ introduced in the last subsection, with respect to this Hermitian structure. To this aim, we first want an integral expression for $\bar{\Psi}$. This could be obtained by just complex conjugating the definition (2.6) of Ψ, but instead, we prefer to introduce a conjugate WZW model describing a map $h: \Sigma \to G$. This time, we introduce a gauge field B gauging the subgroup G_L of $G_L \times G_R$. This is again an anomalous subgroup, so a gauge invariant action $I(h, B)$ extending the WZW action $I(h)$ does not exist. The best that one can do, analogously to (2.3), is

$$I'(h, B) = I(h) - \frac{1}{2\pi} \int_{\Sigma} d^2z\, \mathrm{Tr}\, B_z \partial_{\bar{z}} h \cdot h^{-1} - \frac{1}{4\pi} \int_{\Sigma} d^2z\, \mathrm{Tr}\, B_z B_{\bar{z}}. \tag{2.19}$$

Under the infinitesimal transformation

$$\delta h = uh, \qquad \delta B_i = -D_i u, \tag{2.20}$$

one has

$$\delta I'(h, B) = -\frac{1}{4\pi} \int_{\Sigma} d^2z\, \mathrm{Tr}\, u(\partial_z B_{\bar{z}} - \partial_{\bar{z}} B_z). \tag{2.21}$$

As in (2.6), we now define

$$\begin{aligned} \chi(B) &= \int Dh\, e^{-kI'(h,B)} \\ &= \int Dh \exp\left(-kI(h) + \frac{k}{2\pi} \int_{\Sigma} d^2z\, \mathrm{Tr}\, B_z \partial_{\bar{z}} h \cdot h^{-1} + \frac{k}{4\pi} \int_{\Sigma} d^2z\, \mathrm{Tr}\, B_{\bar{z}} B_z \right). \end{aligned} \tag{2.22}$$

Comparing (2.6) and (2.22), it is evident that in fact χ is the complex conjugate of Ψ, $\chi(A) = \overline{\Psi(A)}$.

We now come to the key step in the present paper. We use these integral representations to compute $|\Psi|^2$:

$$\begin{aligned} |\Psi|^2 &= \frac{1}{\mathrm{vol}(\hat{G})} \int_{\mathscr{A}} DA\, \overline{\Psi(A)}\, \Psi(A) \\ &= \frac{1}{\mathrm{vol}(\hat{G})} \int DgDhDA \exp\left(-kI(g) - kI(h) - \frac{k}{2\pi} \int_{\Sigma} d^2z\, \mathrm{Tr}\, A_{\bar{z}} g^{-1} \partial_z g \right. \\ &\quad \left. + \frac{k}{2\pi} \int_{\Sigma} d^2z\, \mathrm{Tr}\, A_z \partial_{\bar{z}} h \cdot h^{-1} + \frac{k}{2\pi} \int_{\Sigma} d^2z\, \mathrm{Tr}\, A_{\bar{z}} A_z \right). \end{aligned} \tag{2.23}$$

Notice that the integrand is invariant under gauge transformations of the form

$$\delta g = -gu, \quad \delta h = uh, \quad \delta A_i = -D_i u. \tag{2.24}$$

This follows from the cancellation between (2.5) and (2.21).

We can perform the integral over A, using the fact that the exponent in (2.23) is quadratic in A and the operator appearing in the quadratic term is a multiple of the identity.[2] Gaussian integration over A gives

$$|\Psi|^2 = \frac{1}{\mathrm{vol}(\hat{G})} \int DgDh \exp\left(-kI(g) - kI(h) + \frac{k}{2\pi} \int_\Sigma d^2z \,\mathrm{Tr}\, g^{-1}\partial_z g \partial_{\bar z} h \cdot h^{-1} \right). \tag{2.25}$$

At this point we may use a formula of Polyakov and Wiegman [26]:

$$I(gh) = I(g) + I(h) - \frac{1}{2\pi} \int_\Sigma d^2z \,\mathrm{Tr}\, g^{-1}\partial_z g \partial_{\bar z} h \cdot h^{-1}. \tag{2.26}$$

The proof of this formula follows from the following: (i) the formula is obviously valid if $h=1$; (ii) the left- and right-hand sides are both invariant under $h \to wh$, $g \to gw^{-1}$, for arbitrary $w: \Sigma \to G$. To demonstrate (ii), it suffices to check infinitesimal invariance under $\delta g = -gu$, $\delta h = uh$. This can easily be verified directly. Actually, a more conceptual proof of (ii) follows from our above calculation. We know that (2.23) is invariant under (2.24), and integrating out A, an operation that will preserve this symmetry, one deduces that the exponent on the right-hand side of (2.25) has the desired symmetry.

Therefore, replacing the double integral over g and h by a single integral over $f = gh$, and canceling the factor of $\mathrm{vol}(\hat{G})$ in the process, we get

$$|\Psi|^2 = \int Df\, e^{-kI(f)}. \tag{2.27}$$

The right-hand side of (2.27) is by definition the partition function $Z(\Sigma)$ of the WZW model (with symmetry group G and level k) so we have learned

$$Z(\Sigma) = |\Psi|^2, \tag{2.28}$$

which, though still in need of further elucidation, is the statement of holomorphic factorization of the WZW model.

2.3. Varying the Complex Structure of Σ

So far, we have considered the surface Σ with a *fixed* complex structure. At this level, $Z(\Sigma)$ is a number; Ψ is a vector in a fixed vector space V. Equation (2.28) is a relation between them. In this form, the relation is not very remarkable. It gains interest when one permits the complex structure of Σ to vary.

We will work over the space $\mathcal{S}$ of all conformal classes of metrics on Σ. Every conformal metric ϱ determines a complex structure. For any given ϱ, we can define a vector space V_ϱ consisting of holomorphic and gauge invariant sections of the

[2] We can assume a regularization in which the determinant of a multiple of the identity is one. With an arbitrary regularization, such a determinant is a factor of the form $e^{c\chi(\Sigma)}$, where c is a universal constant, independent of Σ, and $\chi(\Sigma)$ is the Euler characteristic of Σ. Such a factor can be removed by adding to the WZW action a multiple of $\int_\Sigma \sqrt{\varrho} R$, where R is the scalar curvature of a metric ϱ that is compatible with the complex structure of Σ

prequantum line bundle $\mathscr{L}^{\otimes k}$ over $\mathscr{A}$. The V_ϱ vary as fibers of a vector bundle $\mathscr{V}$ over $\mathscr{S}$. A section of $\mathscr{V}$ is a function $\Psi(A;\varrho)$ of connections and conformal metrics which, in its dependence on A for fixed ϱ, obeys (2.12) and (2.14).

The space $\mathscr{S}$ is a complex manifold, whose exterior derivative has the standard decomposition $d=\partial+\bar{\partial}$. We will write $\delta^{(1,0)}$ and $\delta^{(0,1)}$ respectively for the ∂ and $\bar{\partial}$ operators of $\mathscr{S}$. One can write these explicitly in the form

$$\begin{aligned}\delta^{(1,0)}&=\int_\Sigma \delta\varrho_{\bar z\bar z}\frac{\delta}{\delta\varrho_{\bar z\bar z}},\\ \delta^{(0,1)}&=\int_\Sigma \delta\varrho_{zz}\frac{\delta}{\delta\varrho_{zz}}.\end{aligned} \tag{2.29}$$

The bundle $\mathscr{V}$ has a (projectively) flat structure, which is defined by giving compatibly a holomorphic structure and an antiholomorphic structure. The holomorphic structure is the "obvious" one. $\Psi(A;\varrho)$ is said to be holomorphic, in its dependence on ϱ, if it is annihilated by

$$V^{(0,1)}=\delta^{(0,1)}. \tag{2.30}$$

For the antiholomorphic structure, we cannot simply use the operator $\delta^{(1,0)}$, since this does not commute with the operator on the left-hand side of (2.12). Rather, as explained in [11, 10], $\Psi(A;\varrho)$ is antiholomorphic if it is annihilated by

$$V^{(1,0)}=\delta^{(1,0)}+\frac{\pi}{2k}\int_\Sigma \delta\varrho_{\bar z\bar z}\,\mathrm{Tr}\frac{D}{DA_{\bar z}}\frac{D}{DA_{\bar z}}. \tag{2.31}$$

It is now just a matter of differentiating under the integral sign to show that $\Psi(A;\varrho)$ as defined in (2.6) is annihilated by $V^{(1,0)}$. This has essentially been done in [13]. We have

$$\delta^{(1,0)}\Psi=\int Dg\,e^{-kI(A,g)}\left(-\frac{k}{8\pi}\int_\Sigma d^2z\delta\varrho_{\bar z\bar z}\varrho^{\bar z z}\,\mathrm{Tr}(g^{-1}D_z g)^2\right), \tag{2.32}$$

where $D_i g=\partial_i g-gA_i$. Similarly,

$$\frac{D}{DA_{\bar z}}\Psi=\int Dg\,e^{-kI(g,A)}\cdot\frac{k}{2\pi}g^{-1}D_z g, \tag{2.33}$$

so that

$$\mathrm{Tr}\frac{D}{DA_{\bar z}}\frac{D}{DA_{\bar z}}\cdot\Psi=\int Dg\,e^{-kI(g,A)}\cdot\left(\frac{k}{2\pi}\right)^2\mathrm{Tr}(g^{-1}D_z g)^2. \tag{2.34}$$

Combining the pieces, we get

$$V^{(1,0)}\Psi=0, \tag{2.35}$$

as was claimed.

Now, let e_α, $\alpha=1,\ldots,\dim V$ be a basis of orthonormal, covariantly constant sections of $\mathscr{V}$ (over some open set in moduli space). Ψ can be expanded in this basis as

$$\Psi(A;\varrho)=\sum_\alpha e_\alpha(A;\varrho)\cdot\overline{f_\alpha(\varrho)} \tag{2.36}$$

with some expansion coefficients $\bar{f}_\alpha$. Equation (2.35) means simply that the $\overline{f_\alpha(\varrho)}$ are anti-holomorphic functions on $\mathscr{S}$ in the usual sense. Consequently, (2.28) amounts to an expression

$$Z(\Sigma;\varrho)=\sum_{\alpha=1}^{\dim V}|f_\alpha|^2 \tag{2.37}$$

for the WZW partition function as a finite sum of norms of holomorphic functions.

The stress tensor of the WZW model is usually defined as

$$T_{zz}=2\frac{\delta}{\varrho_{\bar{z}\bar{z}}}I(g,A)=-\frac{k}{4\pi}\operatorname{Tr}(g^{-1}D_zg)^2\,. \tag{2.38}$$

The current is

$$J_z=\frac{\delta}{\delta A_{\bar{z}}}I(g,A)=\frac{k}{2\pi}g^{-1}D_zg\,. \tag{2.39}$$

The fact that

$$T_{zz}=-(\pi/k)\cdot\operatorname{Tr}J_z^2\,, \tag{2.40}$$

which obviously was the main point in the derivation of (2.35), is known as the (classical form of the) Sugawara-Sommerfield construction. It is well known that when J_z is defined as a quantum operator, $\operatorname{Tr}J_z^2$ must be defined with some point splitting or other regularization; this has the effect of replacing k by $k+h$ (h being the dual Coxeter number of G). See [13, Eq. (49)] for some discussion of this in the present context.

Obviously, our discussion has been purely formal, and we have made no attempt to prove that the key statements, such as the statement (2.28) of holomorphic factorization, survive the quantum anomalies. A proper treatment would have to address the conformal anomalies that affect both Z and Ψ and show that the left- and right-hand sides of (2.28) have the same conformal anomaly and are equal.

Finally, the gauge invariant functional

$$\begin{aligned}I(g,h,A)=I(g,A)+I'(h,A)=I(g)+I(h)+\frac{1}{2\pi}\int_\Sigma d^2z\operatorname{Tr}A_{\bar{z}}g^{-1}\partial_zg\\ -\frac{1}{2\pi}\int_\Sigma d^2z\operatorname{Tr}A_z\partial_{\bar{z}}h\cdot h^{-1}-\frac{1}{2\pi}\int_\Sigma d^2z\operatorname{Tr}A_{\bar{z}}A_z\end{aligned} \tag{2.41}$$

that appeared in the exponent in (2.23) deserves some comment. Let G' be the compact, connected, and simply connected group $G'=G\times G$. The pair $(g,h):\Sigma\to G\times G$ can be regarded as a map of Σ to G'. The G' WZW action is just $I(g,h)=I(g)+I(h)$. Let F be the subgroup of $G'_L\times G'_R$ consisting of elements of the form $((1,a),(a^{-1},1))$. In other words, F acts by $(g,h)\to(ga^{-1},ah)$. Then F is an anomaly free subgroup of $G'_L\times G'_R$ [in the sense that (2.1) is obeyed]. Therefore a gauge invariant action $I(g,h,A)$, reducing to $I(g,h)$ at $A=0$, exists. It is precisely (2.41). Our computation of holomorphic factorization amounted to demonstrating that if $Z_G(\Sigma)$ is the partition function of the WZW model with target G, and $Z_{G'/F}(\Sigma)$ is the partition function of a gauged WZW model with target G' and gauge group F, then

$$Z_G(\Sigma)=Z_{G'/F}(\Sigma)\,. \tag{2.42}$$

Holomorphic factorization has its origin, from this point of view, in the fact that when one computes the action (2.41) of the gauged G'/F model, it turns out to be the *sum* of a functional of g and a functional of h. Since exponentiating the action (to get the integrand of the path integral) turns sums into products, this leads to the ability to factorize $Z_{G'/F}(\Sigma)$ in the fashion that we have described.

3. Holomorphic Factorization of Coset Models

So far we have considered gauged WZW models only as a technical tool in order to understand ordinary WZW models. The gauged WZW models are, however, interesting models of conformal field theory in their own right. For every anomaly-free subgroup F of $G_L \times G_R$ (that is, every subgroup obeying the condition in (2.1)), one has a corresponding gauge invariant generalization of the WZW action, which, upon quantization, leads to a conformal field theory model. The models that arise this way are equivalent to coset models, as has been shown by several authors cited in the introduction.

The most standard examples of anomaly-free subgroups of $G_L \times G_R$ are the following. Let G_{adj} be the diagonal subgroup of $G_L \times G_R$ (acting by $g \to aga^{-1}$, $a \in G$). Let H be any subgroup of G_{adj}. Such an H is always anomaly free.

Let B be an H-valued connection. Since H is an anomaly-free group, a gauge-invariant extension $I(g, B)$ of the WZW action $I(g)$ exists. Explicitly, it is

$$I(g,B)=I(g)-\frac{1}{2\pi}\int_\Sigma d^2z\,\mathrm{Tr}\,B_z\partial_{\bar z}g\cdot g^{-1}+\frac{1}{2\pi}\int_\Sigma d^2z\,\mathrm{Tr}\,B_{\bar z}g^{-1}\partial_z g$$
$$-\frac{1}{2\pi}\int_\Sigma d^2z\,\mathrm{Tr}(B_zB_{\bar z}-B_zgB_{\bar z}g^{-1})\,. \tag{3.1}$$

We want to understand the holomorphic factorization of the corresponding coset model partition function

$$Z_{G/H}(\Sigma)=\frac{1}{\mathrm{vol}(\hat{H})}\int Dg\,DB\,e^{-kI(g,B)}\,. \tag{3.2}$$

This model (with the diagonal embedding of H in $G_L \times G_R$) is sometimes called "the" G/H model, and corresponds to the diagonal modular invariant, as will be clear. In this paper we will only consider these standard anomaly-free subgroups, but the generalization of the considerations to other cases should be apparent.

3.1. *Holomorphic Wave Function*

As in holomorphic factorization of the original WZW model, we now consider a subgroup F of $G_L \times G_R$ which is not anomaly free.[3] In fact, we take $F=H_L \times G_R$, where H_L is the subgroup of G_L coming from the embedding of H in G. An F connection is a pair (B, A), where B and A are H and G connections, respectively. A

[3] The argument could also be expressed in terms of a certain anomaly free subgroup of $G'_L \times G'_R$ where $G'=G\times G$. This formulation would proceed in parallel with the last paragraph of Sect. 2

gauge invariant action $I(g,A,B)$ extending the WZW action does not exist, since the subgroup F of $G_L\times G_R$ is not anomaly free. Analogous to (2.3), there is instead a best possible action, uniquely determined by requiring that the violation of gauge invariance is independent of g and of the conformal structure of Σ. This action is

$$I(g,A,B)=I(g)+\frac{1}{2\pi}\int_\Sigma d^2z\,\mathrm{Tr}\,A_{\bar z}g^{-1}\partial_z g-\frac{1}{2\pi}\int_\Sigma d^2z\,\mathrm{tr}\,B_z\partial_{\bar z}g\cdot g^{-1}$$
$$+\frac{1}{2\pi}\int_\Sigma d^2z\,\mathrm{Tr}\,B_z g A_{\bar z}g^{-1}-\frac{1}{4\pi}\int_\Sigma d^2z\,\mathrm{Tr}(A_zA_{\bar z}+B_zB_{\bar z})\,. \quad (3.3)$$

Under

$$\delta g=vg-gu\,,\qquad \delta A_i=-D_iu\,,\qquad \delta B_i=-D_iv \quad (3.4)$$

(here u and v are zero forms valued, respectively, in the Lie algebras of G and H), we have

$$\delta I(g,A,B)=\frac{1}{4\pi}\int_\Sigma d^2z\,\mathrm{Tr}\,u(\partial_z A_{\bar z}-\partial_{\bar z}A_z-\partial_z B_{\bar z}+\partial_{\bar z}B_z)$$
$$=\frac{i}{4\pi}\int_\Sigma \mathrm{Tr}\,u(dA-dB)\,. \quad (3.5)$$

Before proceeding, let us make a few comments on the global structure. If G has a nontrivial center $Z(G)$, then $Z(G)$, diagonally embedded in $G_L\times G_R$, acts trivially in the WZW model (since $g=aga^{-1}$ for $a\in Z(G)$). The symmetry group that acts faithfully in the WZW model is hence really $(G_L\times G_R)/Z(G)$. Similarly, $F=H_L\times G_R$ does not act faithfully; the group that acts faithfully is $F'=(H_L\times G_R)/Z$, where $Z=H\cap Z(G)$. To make the most precise statements in what follows, it is best to think of the pair (A,B) as a gauge field with structure group F'. The group of maps of Σ to F' will be called $\hat{F}'$. The complexification of F' will be called $F'_{\mathbb{C}}$, and the group of maps of Σ to $F'_{\mathbb{C}}$ will be called $\hat{F}'_{\mathbb{C}}$. The groups of maps of Σ to H, G, and their complexifications $H_{\mathbb{C}}$ and $G_{\mathbb{C}}$ will be called $\hat{H}$, $\hat{G}$, $\hat{H}_{\mathbb{C}}$, and $\hat{G}_{\mathbb{C}}$.

Now, as in Sect. 2, we introduce the holomorphic wave function

$$\chi(A,B)=\int Dg\,e^{-kI(g,A,B)}\,. \quad (3.6)$$

χ obeys certain conditions analogous to those studied in Sect. 2. To exhibit these, we let $\mathscr{A}$ be the space of G-valued connections on Σ, $\mathscr{B}$ the space of H-valued connections, and $\mathscr{C}=\mathscr{A}\times\mathscr{B}$. We want to consider $\mathscr{C}$ as a symplectic manifold, with the symplectic structure given by the formula

$$\omega(a_1,b_1;a_2,b_2)=\frac{1}{2\pi}\int_\Sigma \mathrm{Tr}\,a_1\wedge a_2-\frac{1}{2\pi}\int_\Sigma \mathrm{Tr}\,b_1\wedge b_2\,. \quad (3.7)$$

(Here the a_i and b_j are respectively one forms with values in the Lie algebras of G or H. The pairs (a_1,b_1) and (a_2,b_2) define tangent vectors to $\mathscr{C}$. The "Tr" in the second expression on the right of (3.7) is the quadratic form on the H Lie algebra that is induced from the embedding of H in G.) The minus sign before the second term in (3.7) is characteristic of coset models. Prequantization means constructing a line bundle $\mathscr{L}$ over $\mathscr{C}$ with a unitary connection of curvature $-i\omega$.

Rather as in Sect. 2, the line bundle over $\mathscr{C}$ that is relevant is the trivial line bundle endowed with a connection described by the following formulas:

$$\begin{aligned} \frac{D}{DA_z} &= \frac{\delta}{\delta A_z} - \frac{k}{4\pi} A_{\bar z}, \\ \frac{D}{DA_{\bar z}} &= \frac{\delta}{\delta A_{\bar z}} + \frac{k}{4\pi} A_z, \\ \frac{D}{DB_z} &= \frac{\delta}{\delta B_z} + \frac{k}{4\pi} B_{\bar z}, \\ \frac{D}{DB_{\bar z}} &= \frac{\delta}{\delta B_{\bar z}} - \frac{k}{4\pi} B_z. \end{aligned} \tag{3.8}$$

Computing the curvature of this connection, we see that the trivial line bundle endowed with this connection is isomorphic to $\mathscr{L}^{\otimes k}$, which is how we will refer to it henceforth. The action of the gauge group (that is, the group of maps of Σ to $G \times H$) on $\mathscr{C}$ lifts to an action on $\mathscr{L}^{\otimes k}$. The lift is described at the Lie algebra level by the obvious generalization of (2.17); the G action is generated by the operators

$$D_i \frac{D}{DA_i} - \frac{ik}{4\pi} \varepsilon^{ij} F_{ij}(A), \tag{3.9}$$

and the H action is generated by

$$D_i \frac{D}{DB_i} + \frac{ik}{4\pi} \varepsilon^{ij} F_{ij}(B). \tag{3.10}$$

Here $F(A)$ and $F(B)$ are the curvatures of A and B, respectively.[4]

The analogs of (2.12) and (2.14) are easy to find. χ obeys first of all

$$\frac{D}{DA_z} \chi = \frac{D}{DB_{\bar z}} \chi = 0. \tag{3.11}$$

This has the following interpretation. Pick on $\mathscr{C}$ a complex structure that comes from the standard complex structure on $\mathscr{A}$ and the opposite complex structure on $\mathscr{B}$. (Thus, $A_{\bar z}$ and B_z are holomorphic, and A_z and $B_{\bar z}$ are antiholomorphic.) The (0, 2) part of the curvature of the connection (3.8) vanishes, so $\mathscr{L}^{\otimes k}$ has a natural structure of holomorphic line bundle over $\mathscr{C}$. Equation (3.11) means that χ is a holomorphic section of this line bundle. χ also obeys the analog of (2.14), namely

$$0 = \left(D_i \frac{D}{DA_i} - \frac{ik}{4\pi} \varepsilon^{ij} F_{ij}(A) \right) \chi = \left(D_i \frac{D}{DB_i} + \frac{ik}{4\pi} \varepsilon^{ij} F_{ij}(B) \right) \chi. \tag{3.12}$$

As in the discussion of (2.14), this equation means that χ is gauge invariant in the appropriate sense: it is invariant under the natural lift of the action of the group $\hat{F}'$ of gauge transformations to an action on sections of $\mathscr{L}^{\otimes k}$.

[4] If H (or G) is not connected and simply connected, describing a lift of the gauge group to act on $\mathscr{L}^{\otimes k}$ requires more than the lift of the Lie algebra described by these formulas. The full story is naturally described using the gauge theory approach to prequantization of the space of connections [27–29] and will not be explained here, though the existence of a natural lift is essential later when we consider the role of the center of G

3.2. The Space of Conformal Blocks

Let W be the space of holomorphic sections of $\mathscr{L}^{\otimes k}$ which are $\hat{F}'$ invariant – such as χ. We will devote this subsection to a detailed characterization of W. W is a finite dimensional vector space which can be given the following concrete description. A holomorphic section of $\mathscr{L}^{\otimes k}$ which is $\hat{F}'$ invariant is automatically also $\hat{F}'_{\mathbb{C}}$ invariant. Let $\mathscr{R}=\mathscr{C}/\hat{F}'_{\mathbb{C}}$. According to the Narasimhan-Seshadri theorem, $\mathscr{R}$ is the moduli space of flat F'-valued connections on Σ, up to gauge transformation. $\mathscr{R}$ gets a complex structure from its interpretation as the quotient of the complex manifold $\mathscr{C}$ by the complex group $F'_{\mathbb{C}}$. The holomorphic line bundle $\mathscr{L}^{\otimes k}$ over $\mathscr{C}$ pushes down to a holomorphic line bundle, which we will call by the same name, over $\mathscr{R}$. $\hat{F}'_{\mathbb{C}}$ invariant sections of $\mathscr{L}^{\otimes k}$ over $\mathscr{C}$ are pullbacks of sections of $\mathscr{L}^{\otimes k}$ over $\mathscr{R}$, so $W=H^0(\mathscr{R},\mathscr{L}^{\otimes k})$. This is the space identified in [21, 14] as the space of conformal blocks of the coset model.

W is finite dimensional, since $\mathscr{R}$ is compact. In fact, if Z is trivial, then $\mathscr{R}=\mathscr{M}\times\mathscr{N}$, where $\mathscr{M}=\mathscr{A}/\hat{G}_{\mathbb{C}}$ and $\mathscr{N}=\mathscr{B}/\hat{H}_{\mathbb{C}}$. As is apparent from (3.11), the complex structure on $\mathscr{M}$ is the standard one, and the complex structure on $\mathscr{N}$ is the opposite one. We will refer to $\mathscr{N}$ with the opposite complex structure as $\bar{\mathscr{N}}$. If $\mathscr{L}_{(1)}$ is the standard prequantum line bundle over $\mathscr{M}$ and $\mathscr{L}_{(2)}$ is the standard prequantum line bundle over $\mathscr{N}$ (and we denote their pullbacks to $\mathscr{M}\times\mathscr{N}$ by the same symbols), then $\mathscr{L}^{\otimes k}=\mathscr{L}_{(1)}^{\otimes k}\otimes\mathscr{L}_{(2)}^{\otimes(-k)}$. The minus sign, of course, comes from the minus sign in the second term in (3.7). [As $\mathscr{L}_{(2)}$ has curvature of type (1, 1), it is naturally holomorphic both in the standard complex structure on $\mathscr{N}$ and the opposite one.] Consequently, if Z is trivial,

$$\begin{aligned} W=H^0(\mathscr{R},\mathscr{L}^{\otimes k})&=H^0(\mathscr{M}\times\bar{\mathscr{N}},\mathscr{L}_{(1)}^{\otimes k}\otimes\mathscr{L}_{(2)}^{\otimes(-k)}) \\ &=H^0(\mathscr{M},\mathscr{L}_{(1)}^{\otimes k})\otimes H^0(\bar{\mathscr{N}},\mathscr{L}_{(2)}^{\otimes(-k)}). \end{aligned} \tag{3.13}$$

The space of conformal blocks of the WZW model with target group G, studied in Sect. 2, was

$$V_G=H^0(\mathscr{M},\mathscr{L}_{(1)}^{\otimes k}). \tag{3.14}$$

Likewise, the space of conformal blocks of the WZW model with target group H is

$$V_H=H^0(\mathscr{N},\mathscr{L}_{(2)}^{\otimes k}). \tag{3.15}$$

Here we take $\mathscr{N}$ with its standard complex structure, and a positive tensor power of $\mathscr{L}_{(2)}$. Upon reversing the complex structure on $\mathscr{N}$ and $\mathscr{L}_{(2)}$, we see that, if V_{H^*} is the dual vector space to V_H, then

$$V_{H^*}=H^0(\bar{\mathscr{N}},\mathscr{L}^{\otimes(-k)}). \tag{3.16}$$

Consequently, if Z is trivial,

$$W=V_G\otimes V_{H^*}. \tag{3.17}$$

Now, we want to find the appropriate statement that holds when Z is not trivial. First of all, the natural projection of $F\to F'$ induces a natural map $i:\hat{F}\to\hat{F}'$. i is not quite an embedding; the kernel consists of constant gauge transformations by elements of the center of F. i is also not quite surjective; the quotient $Z'=\hat{F}'/i(\hat{F})$ consists of "twists" by elements of Z in going around closed one-cycles in Σ (described explicitly at the end of this subsection), so in fact $Z'=\mathrm{Hom}(H_1(\Sigma,\mathbb{Z}),Z)$. Thus we have an exact sequence

$$0\to i(\hat{F})\to\hat{F}'\to Z'\to 0. \tag{3.18}$$

Similarly, after complexification (which does not affect finite groups and so leaves Z and Z' unmodified), we have a natural projection $i:\hat{F}_{\mathbb{C}}\to\hat{F}'_{\mathbb{C}}$ and an exact sequence

$$0\to i(\hat{F}_{\mathbb{C}})\to\hat{F}'_{\mathbb{C}}\to Z'\to 0 \tag{3.19}$$

with the same Z'.

We can take the quotient of $\mathscr{C}$ by $\hat{F}'_{\mathbb{C}}$ by first dividing by $i(\hat{F}_{\mathbb{C}})$ and then dividing by Z'. As $\mathscr{C}/i(\hat{F}_{\mathbb{C}})=\mathscr{M}\times\bar{\mathscr{N}}$, we get a natural action of Z' on $\mathscr{M}\times\bar{\mathscr{N}}$, and

$$\mathscr{R}=(\mathscr{M}\times\bar{\mathscr{N}})/Z'. \tag{3.20}$$

From this it follows that, if $X^{Z'}$ denotes the Z' invariant part of a vector space X, then

$$W=(V_G\otimes V_{H*})^{Z'}. \tag{3.21}$$

The Z' action on $\mathscr{M}\times\bar{\mathscr{N}}$ that enters here is easy to describe explicitly. According to the Narasimhan-Seshadri theorem, $\mathscr{M}\times\bar{\mathscr{N}}$ is the moduli space of representations of the fundamental group of Σ in $G\times H$. For Σ of genus r, such a representation is given explicitly by holonomies $(g_1,h_1),\ldots,(g_{2r},h_{2r})$ about $2r$ generating cycles (modulo conjugation, and subject to a well-known relation). Z' acts by $(g_1,h_1),\ldots,(g_{2r},h_{2r})\to(z_1g_1,z_1h_1),\ldots,(z_{2r}g_{2r},z_{2r}h_{2r})$, with $z_1,\ldots,z_{2r}$ being arbitrary elements of Z.

3.3. *Holomorphic Factorization*

The vector space W has a natural Hermitian structure formally given by

$$(\chi_1,\chi_2)=\frac{1}{\mathrm{vol}(\hat{G})\times\mathrm{vol}(\hat{H})}\int_{\mathscr{C}} DA\,DB\,\overline{\chi_1(A,B)}\chi_2(A,B). \tag{3.22}$$

(It is convenient to divide by $\mathrm{vol}(\hat{G})\cdot\mathrm{vol}(\hat{H})$, and not by $\mathrm{vol}(\hat{F}')$, which differs from this by a factor of $\#Z'$, the number of elements in Z'.) We want to show that the partition function of the G/H coset model is

$$Z_{G/H}(\Sigma)=|\chi|^2. \tag{3.23}$$

The reasoning required is very similar to that in Sect. 2, so we will be brief. One first introduces a conjugate WZW model, with gauge group $G_L\times H_R$. The action, for $h:\Sigma\to G$, and A and B gauge fields of G_L and H_R, is

$$\begin{aligned} I'(h,A,B)=I(h)&+\frac{1}{2\pi}\int_\Sigma d^2z\,\mathrm{Tr}\,B_{\bar z}h^{-1}\partial_z h-\frac{1}{2\pi}\int_\Sigma d^2z\,\mathrm{Tr}\,A_z\partial_{\bar z}h\cdot h^{-1}\\ &+\frac{1}{2\pi}\int_\Sigma d^2z\,\mathrm{Tr}\,A_z hB_{\bar z}h^{-1}-\frac{1}{4\pi}\int_\Sigma d^2z\,\mathrm{Tr}(A_zA_{\bar z}+B_zB_{\bar z}). \end{aligned} \tag{3.24}$$

We thus get an integral expression for $\overline{\chi(A,B)}$:

$$\overline{\chi(A,B)}=\int Dh\,e^{-kI'(h,A,B)}. \tag{3.25}$$

Combining (3.6) and (3.25), we get

$$|\chi|^2=\frac{1}{\mathrm{vol}(\hat{G})\cdot\mathrm{Vol}(\hat{H})}\int DA\,DB\,Dg\,Dh\,e^{-kI(g,A,B)-kI'(h,A,B)}. \tag{3.26}$$

Writing out the exponent on the right-hand side of (3.26) explicitly, one sees that it is quadratic in A. The integral over A is Gaussian, therefore. After doing this integral one finds that, using the Polyakov-Wiegman formula, the integral over g and h collapses to an integral over $f=gh$. The remaining functional integral is precisely the definition (3.2) of the partition function $Z_{G/H}(\Sigma)$ of the G/H model, completing the formal proof of (3.23). These steps proceed precisely in parallel with the corresponding points in the derivation of (2.28), and will not be elaborated further.

It remains to consider what happens when the complex structure of Σ varies. Again, the parallel with Sect. 2 is so close that we can be brief. When the complex structure of Σ varies, W varies, as the fiber of a vector bundle $\mathscr{W}$ over the space $\mathscr{S}$ of complex structures on Σ. $\mathscr{W}$ has a projectively flat connection, given by formulas analogous to those of Sect. 2. The holomorphic structure of $\mathscr{W}$ is defined by saying that a section $\chi(A, B; \varrho)$ is holomorphic if it is annihilated by the $(0,1)$ part of the connection

$$\nabla^{(0,1)}=\delta^{(0,1)}-\frac{\pi}{2k}\int_{\Sigma}\delta\varrho_{zz}\operatorname{Tr}\frac{D}{DB_z}\frac{D}{DB_z}. \tag{3.27}$$

The antiholomorphic structure is defined by the $(1,0)$ part of the connection

$$\nabla^{(1,0)}=\delta^{(1,0)}+\frac{\pi}{2k}\int_{\Sigma}\delta\varrho_{\bar{z}\bar{z}}\operatorname{Tr}\frac{D}{DA_{\bar{z}}}\frac{D}{DA_{\bar{z}}}. \tag{3.28}$$

[The justification of these formulas is that ∇ commutes with the operators on the left hand side of (3.11). Alternatively, one can deduce these formulas systematically by working out the Bogoliubov transformation that compensates for a change in polarization of $\mathscr{C}$. The fact that B_z appears in (3.27) and $A_{\bar{z}}$ in (3.28) of course reflects the ubiquitous reversal of sign and change of complex structure of the coset model.] Precisely as in Sect. 2, by differentiating the definition of χ under the integral sign, one shows that χ is antiholomorphic,

$$\nabla^{(1,0)}\chi(A, B; \varrho)=0\,. \tag{3.29}$$

(3.23) and (3.29) make up what is usually called holomorphic factorization of the G/H model.

4. The G/G Model

In this section, we will consider the special case of the G/H coset model for $H=G$. This case is particularly simple, being a topological field theory, and as a result sharper statements can be made. The understanding of these statements also illuminates the "ordinary" models, even the original WZW model, as we will see.

The action of the G/G model is the familiar G/H action,

$$\begin{aligned}I(g, B)=I(g)&-\frac{1}{2\pi}\int_{\Sigma}d^2z\operatorname{Tr}B_z\partial_{\bar{z}}g\cdot g^{-1}+\frac{1}{2\pi}\int_{\Sigma}d^2z\operatorname{Tr}B_{\bar{z}}g^{-1}\partial_z g\\&-\frac{1}{2\pi}\int_{\Sigma}d^2z\operatorname{Tr}(B_zB_{\bar{z}}-B_zgB_{\bar{z}}g^{-1}),\end{aligned} \tag{4.1}$$

specialized to the case $H=G$. Thus, B is now a gauge field valued in the Lie algebra of G. (B is of course gauging the adjoint subgroup of $G_L\times G_R$, so the covariant derivative of g will be $D_i g = \partial_i g + [B_i, g]$.) The main novelty of the case $G = H$ is that this model is a topological field theory, in the sense that (for instance) the partition function

$$Z_{G/G}(\Sigma) = \frac{1}{\mathrm{vol}(\hat{G})} \int DB\, Dg\, e^{-kI(g,B)} \tag{4.2}$$

is independent of the metric of Σ. We will first prove this directly, and then reformulate the argument in the language of holomorphic factorization.

For the direct proof, we note that under an infinitesimal change in the metric of Σ, the action of the gauged WZW model changes according to the following formula:

$$\delta I(g,B) = \frac{1}{8\pi} \int_\Sigma \delta\varrho_{\bar z\bar z}\varrho^{zz} \,\mathrm{Tr}(g^{-1}D_z g)^2 + \frac{1}{8\pi} \int_\Sigma d^2z\, \delta\varrho_{zz}\varrho^{\bar z\bar z} \,\mathrm{Tr}(D_{\bar z} g \cdot g^{-1})^2 . \tag{4.3}$$

The variation of the partition function is hence

$$\delta Z_{G/G}(\Sigma) = \frac{1}{\mathrm{vol}(\hat{G})} \int DB\, Dg\, e^{-kI(g,B)} \cdot \left(-\frac{k}{8\pi}\right) \cdot \int_\Sigma d^2z \left(\delta\varrho_{\bar z\bar z}\varrho^{zz}\,\mathrm{Tr}(g^{-1}D_z g)^2 \right.$$
$$\left. + \int_\Sigma d^2z\, \delta\varrho_{zz}\varrho^{\bar z\bar z}\,\mathrm{Tr}(D_{\bar z} g \cdot g^{-1})^2 \right), \tag{4.4}$$

and we must show that this vanishes. To do this, we will show that the integrand in (4.4) is a total derivative in function space. In fact, since the variation of the action in a change of the connection B is

$$\delta I(g,B) = \frac{1}{2\pi} \int d^2z\, \mathrm{Tr}\, \delta B_{\bar z} g^{-1} D_z g - \frac{1}{2\pi} \int_\Sigma d^2z\, \mathrm{Tr}\, \delta B_z D_{\bar z} g \cdot g^{-1}, \tag{4.5}$$

we get

$$\int DB\, Dg \int_\Sigma d^2z\, \delta\varrho_{\bar z\bar z}\varrho^{zz} \frac{\delta}{\delta B^a_{\bar z}} \cdot ((g^{-1}D_z g)^a \cdot e^{-kI(g,B)})$$
$$= \int DB\, Dg\, e^{-kI(g,B)} \cdot \left(-\frac{k}{2\pi}\right) \cdot \int_\Sigma d^2z\, \delta\varrho_{\bar z\bar z}\varrho^{zz}\,\mathrm{Tr}(g^{-1}D_z g)^2 . \tag{4.6}$$

Assuming that one can integrate by part in function space, the left-hand side of (4.6) vanishes, and this means that the first term on the right-hand side of (4.4) can be discarded. The second term on the right-hand side of (4.4) similarly vanishes since

$$\int DB\, Dg \int_\Sigma d^2z\, \delta\varrho_{zz}\varrho^{\bar z\bar z} \frac{\delta}{\delta B^a_z} \cdot ((D_{\bar z} g \cdot g^{-1})^a \cdot e^{-kI(g,B)})$$
$$= \int DB\, Dg\, e^{-kI(g,B)} \cdot \left(\frac{k}{2\pi}\right) \cdot \int_\Sigma d^2z\, \delta\varrho_{zz}\varrho^{\bar z\bar z}\,\mathrm{Tr}(D_{\bar z} g \cdot g^{-1})^2 . \tag{4.7}$$

4.1. *Factorization*

The attentive reader will note that the key fact in the last paragraph was the Sugawara-Sommerfield construction (2.40), which also played a key role in the analysis of holomorphic factorization of general WZW and coset models. In fact, it is illuminating to recast the above argument in the language of holomorphic factorization.

Precisely as in the general discussion of coset models, we consider the gauging of $H_L \times G_R$ (which now is $G_L \times G_R$). The closest to a gauge invariant action is (3.17), which we repeat for convenience:

$$I(g, A, B) = I(g) + \frac{1}{2\pi} \int_\Sigma d^2z \operatorname{Tr} A_{\bar z} g^{-1} \partial_z g - \frac{1}{2\pi} \int_\Sigma d^2z \operatorname{Tr} B_z \partial_{\bar z} g \cdot g^{-1}$$
$$+ \frac{1}{2\pi} \int_\Sigma d^2z \operatorname{Tr} B_z g A_{\bar z} g^{-1} - \frac{1}{4\pi} \int_\Sigma d^2z \operatorname{Tr}(A_z A_{\bar z} + B_z B_{\bar z}). \quad (4.8)$$

The novelty, compared to the case of arbitrary H, is that now there is a kind of symmetry between the G_L and G_R gauge fields B and A: (4.8) is invariant under reversing the complex structure (or equivalently, the orientation) of Σ, exchanging g with g^{-1}, and exchanging A and B. [Alternatively, if one exchanges g with g^{-1} and exchanges A and B, while leaving the orientation of Σ fixed, then (4.8) is complex conjugated.] We want to see the consequences of this.

Just as in the general case of arbitrary H, one introduces the holomorphic wave function

$$\chi(A, B) = \int Dg\, e^{-kI(g, A, B)}. \quad (4.9)$$

The general arguments specialized to this case show that the norm of χ is

$$|\chi|^2 = Z_{G/G}(\Sigma) \quad (4.10)$$

and that χ is anti-holomorphic,

$$\nabla^{(1,0)}\chi = 0. \quad (4.11)$$

The novelty is the symmetry between A and B, which reverses the complex structures, and so makes it apparent that χ must also be holomorphic,

$$\nabla^{(0,1)}\chi = 0. \quad (4.12)$$

Equations (4.11) and (4.12) can both be proved by using the general definition (3.27) and (3.28) of the connection and differentiating under the integral sign, as in the proof of (2.35).

Equations (4.11) and (4.12) together mean that χ is covariantly constant, and hence $|\chi|^2$ is a constant. From the factorization law (4.10) we thus deduce again that $Z_{G/G}(\Sigma)$ is independent of the metric.

To probe more deeply, we now recall the general description in Sect. 3 of the vector bundle $\mathscr{W}$ in which holomorphic factorization of the coset model takes place. We had

$$\chi \in (V_G \otimes V_{H^*})^{Z'}, \quad (4.13)$$

with

$$V_G = H^0(\mathscr{M}, \mathscr{L}^{\otimes k}),$$
$$V_H = H^0(\mathscr{N}, \mathscr{L}^{\otimes k}). \quad (4.14)$$

Setting $H = G$ and interpreting $V_G \otimes V_{G^*}$ as $\operatorname{Hom}(V_G, V_G)$, we have

$$\chi \in (\operatorname{Hom}(V_G, V_G))^{Z'}. \quad (4.15)$$

Now, $\operatorname{Hom}(V_G, V_G)$ contains a canonical (and Z'-invariant) element, the identity map 1; it is natural to ask whether $\chi = 1$.

The "symmetry" between A and B makes it obvious that χ is hermitian (in the natural norm on V_G). Indeed $\chi(A,B)=\overline{\chi(B,A)}$ because (4.8) is complex conjugated if one exchanges A and B while mapping $g\to g^{-1}$. With our methods it is also easy to prove that

$$\chi^2=\chi\,. \tag{4.16}$$

This amounts to the statement that

$$\chi(A,C)=\frac{1}{\operatorname{vol}(\hat{G})}\int DB\,\chi(A,B)\chi(B,C)\,. \tag{4.17}$$

This is proved by replacing each of the three copies of χ that appear in (4.17) with the integral representation (4.9), performing the Gaussian integral over B, and using the Polyakov-Wiegmann formula (just as in the original proof of holomorphic factorization in Sect. 2.2).

Equation (4.16) means that χ is the orthogonal projection operator onto a subbundle $\mathscr{V}'$ of $\mathscr{V}$ (whose inclusion in $\mathscr{V}$ is compatible with the projectively flat connection on $\mathscr{V}$, since χ is covariantly constant). It will be evident presently that holomorphic factorization can be carried out in $\mathscr{V}'$. One expects that $\chi=1$ and $\mathscr{V}'=\mathscr{V}$, but the methods of this paper do not seem to suffice for proving this.

The fact that χ is covariantly constant means that if $e_i(A;\varrho)$ is an orthonormal basis of covariantly constant sections of $\mathscr{V}'$, then $\chi(A,B;\varrho)=\sum_{i,j}Q_{ij}e_i(A;\varrho)\overline{e_j(B;\varrho)}$, with some *constants* $Q_{i,j}$. The fact that $\chi^2=\chi$ (and $\chi=1$ when restricted to $\mathscr{V}'$, by definition of $\mathscr{V}'$) means that $Q_{i,j}=\delta_{ij}$. So

$$\chi(A,B;\varrho)=\sum_{i=1}^{\dim\mathscr{V}'}e_i(A;\varrho)\overline{e_i(B;\varrho)}\,. \tag{4.18}$$

We can thus compute the norm of χ to get

$$Z_{G/G}(\Sigma)=|\chi|^2=\dim(\mathscr{V}')\,. \tag{4.19}$$

One expects that $\mathscr{V}=\mathscr{V}'$, but in any case, if this is not true, it is $\mathscr{V}'$ that should be called the space of conformal blocks in the WZW model. (This will be even more apparent in the next subsection.) So we have established that the partition function of the G/G model is the number of conformal blocks of the WZW model, a result that has been conjectured by Spiegelglas [22], with considerable evidence.

4.2. *Relation to the* WZW *Model and "Ordinary" Coset Models*

Now we will see what we can learn about the original WZW model, and general coset models, by applying our knowledge of the G/G model. The reason that one can learn something interesting is that, upon returning to the definition (4.9) of χ, and noting that $I(g,0,0)$ is the original action of the WZW model, we see that the partition function of the WZW model is

$$Z_G(\Sigma)=\chi(0,0;\varrho)\,. \tag{4.20}$$

In view of (4.18), we get therefore

$$Z_G(\Sigma)=\sum_{i=1}^{\dim\mathscr{V}'}e_i(0;\varrho)\overline{e_i(0;\varrho)}\,. \tag{4.21}$$

This formula expresses the partition function of the WZW model in terms of quantities that naturally arise in quantizing the moduli space $\mathscr{A}$ of G-valued connections, namely the orthonormal parallel sections $e_i(A;\varrho)$.

As a check, let us verify that (4.21) is compatible with the earlier description of $Z_G(\Sigma)$ as the norm squared of a holomorphic section of $\mathscr{V}$:

$$Z_G(\Sigma)=|\Psi|^2 . \tag{4.22}$$

Recalling the definition (2.6) of Ψ, we see that $\Psi(A;\varrho)=\chi(A,0;\varrho)$, so from (4.18) we get

$$\Psi(A;\varrho)=\sum_{i=1}^{\dim \mathscr{V}'} e_i(A;\varrho)\overline{e_i(0;\varrho)} . \tag{4.23}$$

As the e_i are orthonormal, insertion of this in (4.22) gives back (4.21).

In a similar fashion, one can also obtain a formula for the partition function of the G/H model. Recalling the definitions (4.14) of V_G and V_H, we see that there is a natural map $r_{G/H}: V_G \to V_H$, which takes a section of $\mathscr{L}^{\otimes k}$ over $\mathscr{M}$ and restricts it to $\mathscr{N}$. (As $\mathscr{M}$ and $\mathscr{N}$ are the moduli spaces of holomorphic $G_{\mathbb{C}}$ and $H_{\mathbb{C}}$ bundles, respectively, there is a natural inclusion of $\mathscr{N}$ in $\mathscr{M}$.) Taking complex conjugates, there is also a natural map $r^*_{G/H}: V^*_G \to {}^H$. These maps do not respect the unitary structures.

For every H, we have holomorphic factorization

$$Z_{G/H}(\Sigma)=|\chi_{G/H}(A,B;\varrho)|^2 , \tag{4.24}$$

where $\chi_{G/H}$ is the functional defined in (3.6). Inspecting the definition, we see that

$$\chi_{G/H}(A,B;\varrho)=(1\otimes r^*_{G/H})\chi_{G/G}(A,B;\varrho) . \tag{4.25}$$

Using (4.18) we can now rewrite (4.24) in the form

$$Z_{G/H}(\Sigma)=\sum_{i=1}^{\dim \mathscr{V}_G} |r_{G/H}(e_i)|^2 . \tag{4.26}$$

Alternatively,

$$Z_{G/H}(\Sigma)=\sum_{i=1}^{\dim \mathscr{V}_G}\sum_{j=1}^{\dim V_H} |(f_j, r_{G/H}e_i)|^2 , \tag{4.27}$$

with f_j an orthonormal basis of parallel sections of $\mathscr{V}_H$. Formulas of this type were suggested in [13, 14]. If one takes H to be the trivial group (with only the identity element), then (4.26) reduces, as it should, to (4.21).

Appendix

The purpose of this appendix is to clarify the geometric meaning of the classical gauged WZW actions on which this paper is based. Some readers may wish to consult this appendix before reading the body of the paper (see also [4, 29, 30]).

The problem can be clarified by formulating it in more generality than we actually need. We consider an arbitrary connected manifold M with a closed three-form ω whose periods are multiples of 2π, so that ω is related to a class in $H^3(M,\mathbb{Z})$. We let Σ be an oriented two dimensional surface without boundary. To simplify the considerations that follow, we assume that $\pi_1(M)=\pi_2(M)=0$, so that a continuous map $X:\Sigma\to M$ is automatically nullhomotopic. (The main novelty that arises if one relaxes this condition is that one must use integral cohomology instead of just working with differential forms.) We suppose given the action of a compact Lie group F on M and we suppose that ω is F invariant. To simplify the story, we suppose that F is simple and simply-connected. (Again, if these conditions are relaxed, the main novelty that arises is that one must use integral F-equivariant cohomology, rather than the de Rham model that will appear presently.) We describe the Lie algebra of F with generators T_a and relations

$$[T_a, T_b]=f_{ab}^c T_c. \tag{A.1}$$

Let

$$\Gamma(X)=\int_B X^*\omega, \tag{A.2}$$

where B is any three manifold with $\partial B=\Sigma$, and an arbitrary extension of X over B has been chosen. Γ has values in $\mathbb{R}/2\pi\mathbb{Z}$. We wish to construct a gauge invariant generalization of Γ.

The action of F on M is generated by vector fields V_a. Introducing a gauge field $A=\sum_a A^a T_a$, with structure group F, we want to find a generalization $\Gamma(X,A)$ of Γ that is invariant under

$$\delta X=\varepsilon^a V_a, \qquad \delta A^a=-D\varepsilon^a, \tag{A.3}$$

for ε^a an infinitesimal gauge transformation. The variation of Γ is

$$\delta\Gamma=\int_\Sigma \varepsilon^a X^*(i_{V_a}(\omega)). \tag{A.4}$$

(i_V is the operation of contracting with a vector field V.) Additional terms that can be added to (A.2) to cancel this exist only if there are one-forms λ_a on M such that

$$i_{V_a}(\omega)=d\lambda_a, \tag{A.5}$$

and moreover such that

$$i_{V_a}(\lambda_b)+i_{V_b}(\lambda_a)=0. \tag{A.6}$$

If such λ_a exist, then, by averaging suitably over the compact group F, one can suppose that they transform in the adjoint representation of F. In this case, the desired gauge invariant generalization of Γ is

$$\Gamma(X,A)=\Gamma(X)-\int_\Sigma A^a\wedge X^*(\lambda_a)-\frac{1}{2}\int_\Sigma A^a\wedge A^b\cdot X^*(i_{V_b}\lambda_a). \tag{A.7}$$

Equations (A.5) and (A.6) have a geometrical meaning, in terms of the so-called F-equivariant cohomology of M, denoted $H_F^*(M)$. A de Rham model for this equivariant cohomology, explained in [31, 32], can be described as follows. Let $\Omega^*(M)$ be the de Rham complex of M, and let $S^*(\mathscr{F})$ be a symmetric algebra on the Lie algebra $\mathscr{F}$ of F, with generators ϕ^a considered to be of degree two. Let $W^*=(\Omega^*(M)\otimes S^*(\mathscr{F}))^F$ (with ^{F} denoting the F invariant part). In W^*, introduce the differential

$$D=d+\sum_a \phi^a i_{V_a}. \tag{A.8}$$

If ω is a closed form on M, an element $\bar{\omega} \in W^*$ is called an equivariant extension of ω if $D\bar{\omega}=0$ and $\bar{\omega}|_{\phi=0}=\omega$. The meaning of (A.5) and (A.6) is simply that they are the conditions for ω to have an equivariant extension. In fact,

$$\bar{\omega}=\omega-\sum_a \phi^a \lambda_a \tag{A.9}$$

is an equivariant extension of ω if and only if the λ_a obey (A.5) and (A.6) and transform in the adjoint representation of F.

Now let us specialize to the case of actual interest in this paper in which M is the group manifold of a simple, compact, connected, and simply-connected Lie group G, and

$$\omega=\frac{1}{12\pi}\operatorname{Tr}(g^{-1}dg)^3 . \tag{A.10}$$

Moreover, F is a connected subgroup of $G_L \times G_R$. The embedding of F in G_L and G_R is determined by an embedding of Lie algebras which we can write as

$$T_a \to (T_{a,L}, T_{a,R}) . \tag{A.11}$$

The vector fields V_a are described by the formula

$$\delta g=\sum_a \varepsilon^a (T_{a,L}g - gT_{a,R}) . \tag{A.12}$$

One has

$$i_{V_a}\omega = d\lambda_a \tag{A.13}$$

with

$$\lambda_a=\frac{1}{4\pi}\operatorname{Tr}(T_{a,L}(dg\cdot g^{-1})+T_{a,R}(g^{-1}dg)) . \tag{A.14}$$

These λ_a transform in the adjoint representation of F. The non-uniqueness in the choice of the λ_a is $\lambda_a \to \lambda_a + dw_a$, where the w_a are zero forms in the adjoint representation of F. Equation (A.14) is the unique universal formula that works for any F. One now computes that

$$i_{V_a}(\lambda_b)+i_{V_b}(\lambda_a)=\frac{1}{2\pi}\operatorname{Tr}(T_{a,L}T_{b,L}-T_{a,R}T_{b,R}) . \tag{A.15}$$

[Note that the possible w_a do not contribute since $i_{V_a}(dw_b)=f_{ab}^c w_c$ is antisymmetric in a and b.] Thus the equivariant extension $\bar{\omega}$ of ω and the corresponding gauge invariant extension $\Gamma(g, A)$ of Γ exist precisely if F is such that the right-hand side of (A.15) vanishes.

This is the criterion that was stated in (2.1). The gauge invariant extension of Γ, when F is such that (A.15) vanishes, is explicitly

$$\begin{aligned}\Gamma(g, A)=\Gamma(g)-\frac{1}{4\pi}\int_\Sigma A^a \wedge \operatorname{Tr}(T_{a,L}dg\cdot g^{-1}+T_{a,R}g^{-1}dg) .\\ -\frac{1}{8\pi}\int A^a \wedge A^b \operatorname{Tr}(T_{a,R}g^{-1}T_{b,L}g-T_{b,R}g^{-1}T_{a,L}g) .\end{aligned} \tag{A.16}$$

Even when (A.15) does not vanish, (A.16) is the closest that there is to a gauge invariant extension of $\Gamma(g, A)$, in the sense that the variation of (A.16) under a gauge transformation depends only on A and is independent of g. This fact, which played an important role in the body of the paper, reflects the fact that the λ_a obeying (A.5) exist for any F; only the validity of (A.6) depends on F. This means that although an extension $\bar{\omega}$ of ω obeying $D\bar{\omega}=0$ may not exist, ω always has an extension such that

$$D\bar{\omega}\in S^*(\mathscr{F}). \tag{A.17}$$

(And such an $\bar{\omega}$ is unique if one wishes a formula that works universally for any F.) This relation precisely ensures that the violation of gauge invariance depends on A and not g.

Geometrically, the reason that (A.17) has a solution is as follows. The equivariant cohomology of G is the cohomology of the homotopy quotient $G//F = G\times_F EF$. If one computes the cohomology of $G//F$ from the spectral sequence of the fibration $G//F\to BF$, one sees (since ω is a three dimensional class, and the non-trivial cohomology of BF begins in dimension four) that the only obstruction to existence of an equivariant extension $\bar{\omega}$ of ω comes from $H^4(BF)$. In fact, the invariant quadratic form on the F Lie algebra that appears on the right-hand side of (A.15) represents the obstruction class in $H^4(BF)$, via the Chern-Weil homomorphism. The cohomology of BF is isomorphic to $S^*(\mathscr{F})$, so the obstruction is an element of $S^*(\mathscr{F})$.

The gauge invariant generalization of the WZW Lagrangian is

$$I(g,A)=-\frac{1}{8\pi}\int_\Sigma \operatorname{Tr} g^{-1}d_A g\wedge * g^{-1}d_A g - i\Gamma(g,A), \tag{A.18}$$

with $*$ the Hodge star operator, d_A the gauge-covariant extension of the exterior derivative, and $\Gamma(g,A)$ given in (A.16). The first term depends on the conformal structure of Σ, and the second has a topological origin that we have attempted to elucidate in this appendix. The properties of the WZW model depend on a peculiar interplay between the two terms, some aspects of which we have seen in this paper. All the particular formulas for gauged WZW Lagrangians given in this paper are various specializations of (A.18).

References

1. Witten, E.: Non-Abelian bosonization in two dimensions. Commun. Math. Phys. **92**, 455 (1984)
2. Wess, J., Zumino, B.: Consequences of anomalous ward identities. Phys. Lett. **37** B, 95 (1971)
3. Witten, E.: Global aspects of current algebra. Nucl. Phys. B **223**, 422 (1983)
4. Felder, G., Gawedzki, K., Kupianen, A.: Spectra of Wess-Zumino-Witten models with arbitrary simple groups. Commun. Math. Phys. **117**, 127 (1988)
   Gawedzki, K.: Topological actions in two-dimensional quantum field theories. In: Non-perturbative quantum field theory. 't Hooft, G. et al. (eds.). London: Plenum Press 1988
5. Knizhnik, V.G., Zamolodchikov, A.B.: Current algebra and Wess-Zumino model in two dimensions. Nucl. Phys. B **247**, 83 (1984)
6. Friedan, D., Shenker, S.: The analytic geometry of two-dimensional conformal field theory. Nucl. Phys. B **281**, 509–545 (1987)
7. Verlinde, E.: Fusion rules and modular transformations in 2d conformal field theory. Nucl. Phys. B **300**, 351
8. Moore, G., Seiberg, N.: Polynomial equations for rational conformal field theories. Phys. Lett. B **212**, 360 (1988); Classical and quantum conformal field theory. Nucl. Phys. B
9. Witten, E.: Quantum field theory and the Jones polynomial. Commun. Math. Phys. **121**, 351 (1989)
10. Elitzur, S., Moore, G., Schwimmer, A., Seiberg, N.: Remarks on the canonical quantization of the Chern-Simons-Witten theory. IAS preprint HEP-89/20

11. Axelrod, S., DellaPietra, S., Witten, E.: Geometric quantization of Chern-Simons gauge theory. J. Diff. Geom. **33**, 787 (1991)
12. Hitchin, N.: Flat connections and geometric quantization. Commun. Math. Phys. **131**, 347 (1990)
13. Gawedzki, K.: Constructive conformal field theory. In: Functional integration, geometry, and strings. Hava, Z., Sobczyk, J. (eds.). Boston, Basel: Birkhäuser 1989
14. Gawedzki, K., Kupianen, A.: G/H conformal field theory from gauged WZW model. Phys. Lett. **215**B, 119 (1988); Coset construction from functional integrals. Nucl. Phys. B **320 (FS)**, 649 (1989)
15. Bernard, D.: On the Wess-Zumino-Witten models on the torus. Nucl. Phys. B **303**, 77 (1988); On the Wess-Zumino-Witten models on Riemann surfaces. Nucl. Phys. B **309**, 145 (1988)
16. Guadagnini, E., Martellini, M., Minchev, M.: Phys. Lett. **191**B, 69 (1987)
17. Bardacki, K., Rabinovici, E., Saring, B.: Nucl. Phys. B **299**, 157 (1988)
Altschuler, A., Bardacki, K., Rabinovici, E.: Commun. Math. Phys. **118**, 241 (1988)
18. Karabali, D., Park, Q.-H., Schnitzer, H.J., Yang, Z.: Phys. Lett. **216B**, 307 (1989)
Schnitzer, H.J.: Nucl. Phys. B **324**, 412 (1989)
Karabali, D., Schnitzer, H.J.: Nucl. Phys. B **329**, 649 (1990)
19. Goddard, P., Kent, A., Olive, D.: Phys. Lett. B **152**, 88 (1985)
20. Bardacki, K., Halpern, M.B.: Phys. Rev. D **3**, 2493 (1971)
Halpern, M.B.: Phys. Rev. D **4**, 2398 (1971)
21. Moore, G., Seiberg, N.: Taming the conformal zoo. Phys. Lett. B
22. Spiegelglas, M.: Lecture at IAS (October, 1990), Setting Fusion Rules in Topological Landau-Ginzburg. Technion preprint; Spiegelglas, M., Yankielowicz, S.: G/G Topological Field Theory by Cosetting. Fusion Rules As Amplitudes in G/G Theories. Preprints (to appear)
23. Kostant, B.: Orbits, symplectic structures, and representation theory, Proc. of the U.S.-Japan Seminar in Differential Geometry (Kyoto, 1965); Quantization and Unitary Representations. Lecture Notes in Math., vol. 1170, p. 87. Berlin, Heidelberg, New York: Springer 1970; Line Bundles and the Prequantized Schrodinger Equation. Coll. Group Theoretical Methods in Physics (Marseille, 1972) p. 81
24. Souriau, J.: Quantification geometrique. Commun. Math. Phys. **1**, 374 (1966). Structures des systemes dynamiques. Paris: Dunod 1970
25. Gawedzki, K.: Quadrature of conformal field theories. Nucl. Phys. **328**, 733 (1989)
26. Polyakov, A.M., Wiegman, P.B.: Theory of non-abelian Goldstone bosons in two dimensions. Phys. Lett. B **131**, 121 (1983)
27. Ramadas, T.R., Singer, I.M., Weitsman, J.: Some comments on Chern-Simons gauge theory. MIT preprint
28. Freed, D.: Preprint (to appear)
29. Axelrod, S.: Ph. D. thesis, Princeton University (1991), Chapter four
30. Hull, C.M., Spence, B.: The geometry of the gauged sigma model with Wess-Zumino Term. Queen Mary and Westfield College preprint QMW 90/04
31. Atiyah, M.F., Bott, R.: The moment map and equivariant cohomology. Topology **23**, 1 (1984)
32. Mathai, V., Quillen, D.: Superconnections, Thom classes, and equivariant differential forms. Topology **25**, 85 (1986)

Communicated by A. Jaffe

Bibliography

[Atiyah88] M. Atiyah, "New invariants of 3- and 4-dimensional manifolds", *Proc. of Sym. in Pure Math.*, **48**(1988), 285-299.

[Atiyah] M. Atiyah, *The geometry and physics of Knots*, Cambridge University Press, 1990.

[Atiyah-Bott(1982)] M. Atiyah and R. Bott, "The Yang-Mills equation over Riemann surfaces", *Philo. Trans. Royal Soc. London* **A308**(1982).

[Axelrod-DellaPietra-Witten] S. Axelrod, S. DellaPietra and E. Witten, "Geometric quantization of Chern-Simons gauge theory", *J. of Diff. Geom.* **33**(1991) 787-902.

[Axelrod-Singer] S. Axelrod and I. Singer, "Chern-Simons perturbation theory II", *J. of Diff. Geom.* **39**(1994) 173-213.

[Balavin-Polyakov-Zamalodchikov] A. Balavin, A. Polyakov, A. Zamalodschikov, "Infinite conformal symmetry in two-dimensional quantum field theory", *Nucl. Phys.* **B241**(1984) 333-380.

[Bar-Natan-Witten] Dror Bar-Natan and E. Witten, "Perturbative expansion of Chern-Simons theory with non-compact gauge group", *Comm. Math. Phys.* **141**(1991) 423-440.

[Bar-Natan] Dror Bar Natan, "On the Vassiliev knot invariant" *Topology***34**(1995) 423-472.

[Birman and Lin] Joan Birman and Xiao-Song Lin, "Knot polynomials and Vassiliev invariant" *Inventione Math.* **111**(1993) 225-270.

[Blasi and Collina] A. Blasi and R. Collina, "Finiteness of the Chern-Simons model in perturbation theory", *Nucl. Phys.***B345**(1990) 472-492.

[Bott] R. Bott, *Michigan Journal of Mathematics***14**(1967) 231-244.

[Chen] K. T. Chen, "Iterated path integrals", *Bull. of AMS* **83**(1977) 831-879.

[Chen-Semenoff-Wu] W. Chen, G. Semenoff and Y. S. Wu, "Two-loop analysis of non-Abelian Chern-Simons theory", *Phys. Rev.* **46**(1992) 5521.

[Chern-Simons] S. S. Chern and J. Simons, "Characteristic forms and geometric invariants", *Ann. Math.***99**(1974) 48-69.

[Chern] S. S. Chern, "On a conformal invariants of three-dimensional manifolds", *Aspects of Mathematics and Its Applications*, Elsevier Science Publishers B. V. (1986),pp 245-252; S. S. Chern: Selected Papers, **Vol. 4** (1989) Springer 181-188.

[Donaldson] S. K. Donaldson, "Remarks on gauge theory, complex geometry and 4-manifold topology", *Fields Medallists Lectures*, edited by M. Atiyah *et al.*, World Scientific, 1997.

[Faddeev] L. D. Faddeev, *Gauge Fields, Introduction to Quantum Theory*, Addison-Wesley, 1991.

[Jeffrey-Kirwan] L. C. Jeffrey and F. C. Kirwan, "Localization and quantization conjecture", *Topology* **36**(1997) 647-693.

[Kirwan] F. Kirwan, "Cohomology of quotients in symplectic and algebraic geometry", *Mathematical Notes 31*, Princeton University Press, 1984.

[Hitchin] N. Hitchin, "Flat connections and geometric quantization", *Comm. Math. Phys.* **131** (1990) 347-380.

[Kontsevich] M. Kontsevich, "Rational conformal field theory and invariants of 3-dimensional manifolds".

[Kauffman] Luis Kauffman, "Knots and Physics", World Scientific.

[Knapp-Vogan] A. Knapp and D. Vogan, "Cohomological Induction and Unitary Representations", Princeton University Press, 1995.

[Kohno] T. Kohno, "Topological invaraints for 3-manifolds using representations of mapping class groups, I", *Topology* **31**(1992) 203-230.

[Meinrenken] E. Meinrenken, "Symplectic surgery and the $Spin^c$−Dirac operator, *Ad. Math.* 134(1998) 240-277.

[Moore and Seiberg] G. Moore and N. Seiberg, "Classical and quantum conformal field theory", *Comm. Math. Phys.***123**(1989) 177-254.

[Narasimhan-Seshadri] M. S. Narasimhan and C. S. Seshadri, "Stable and unitary vector bundles on a compact Riemann surface", *Ann. Math.* **82**(1965) 540-567.

[Reshetikhin-Turaev] N. Yu. Reshetikhin, V. G. Turaev, "Invariants of three-manifolds via link polynomials and quantum groups".

[Quillen] D. G. Quillen, "Determinants of Cauchy-Riemann operators over a Riemann surface", *Func. Anal. Appl.* **19**(1986) 31.

[Sawin] S. Sawin, "Links, quantum groups and TQFTS", *Bull. of AMS* **33**(1996) 413-445.

[Schwarz] A. Schwarz, "The partition function of degenerate quadratic functionals and Ray-Singer invariants", *Lett. Math. Phys.* **2**(1978) 247.

[Tian-Zhang] Y. Tian and W. Zhang, "An analytic proof of the geometric quantization conjecture of Guillemin-Sternberg", *Invent. Math.* **132**(1998) 229-259.

[Tsuchiya-Kanie] A. Tsuchiya and Y. Kanie, "Vertex operators in conformal field theory on $\mathbf{P}^1$ and monodromy representations of braid groups", *Advanced Studies in Pure Math.* **16** (1988) 297-372, (Errata) ibid **19**(1990) 675-682.

[Uhlenbeck-Yau] K. Uhlenbeck and S. T. Yau, "On the existence of Hermitian-

Yang-Mills connections in stable vector bundles", *Comm. Pure Appl. Math.* Vol **34**(1986) 257-293.

[Vafa] C. Vafa, "On the Gauge Theory/Geometry Correspondence", hep-th/9811131.

[Witten-sigma] E. Witten, "Topological Sigma Models", *Comm. Math. Phys.*.**118**(1988) 411-449.

[Witten] E. Witten, "Quantum field theory and the Jones polynomials", *Comm. Math. Phys.* **121**(1989) 351-399.

[Witten-Casson] E. Witten, "Topology-changing amplitudes in $2+1$ dimensional gravity", *Nucl. Phys.* **B323**(1989) 113-140.

[Witten-quantum-group] E. Witten, "Gauge theories, vertex models, and quantum groups", *Nucl. Phys.* **B330**(1990) 285-346.

[Witten-localization] E. Witten, "Two dimensional gauge theories revisited".

[Witten-non-compact] E. Witten, "Quantization of Chern-Simons theory with complex gauge group", *Comm. Math. Phys.* **137**(1991) 29-66.

[Witten-combinatorics] E. Witten, "On quantum gauge theories in two dimensions", *Comm. Math. Phys.* **141**(1991) 153-209.

[Witten-string] E. Witten, "Chern-Simons gauge theory as a string theory", *The Floer memorial volume, Prog. Math.* **133**, Birkhauser, Basel, 1995.

[Liu1] Kefeng Liu, "Heat kernel and moduli spaces I", *Math. Research Letter* **3**, (1996) 743–762.

[Liu2] Kefeng Liu, "Heat kernel and moduli space II", *Math. Research Letter* **4**, (1997) 569–588.

[illegible] connections on stable vector bundles", Comm. Pure Appl. Math. Vol. 39 (1986) [illegible]

[Vafa] C. Vafa, [illegible] Geometry", [illegible] hep-th/9811181.

[illegible] and E. Witten, [illegible] Models", [illegible] Math. Phys. 118 (1988) 411–449.

[Witten] E. Witten, "Quantum field theory and the Jones polynomial", Comm. Math. Phys. 121 (1989) 351–399.

[illegible]

[illegible]

[illegible]

[illegible]

Index

Afterwards

This monograph arises from E. Witten's lectures on topological quantum field theory in the spring of 1989 at Fine Hall of Princeton University. At that time E. Witten unified several important mathematical works in terms of quantum field theory, most notably, Donaldson polynomial, Gromov/ Floer homology and Jones polynomials.

In his lectures Witten explained his three-dimensional intrinsic construction of Jones polynomials via Chern-Simons gauge theory. His construction leads to many beautiful applications such as skein relation, surgery formula and a proof of Verlinde's formula. The heart of the construction is to quantize Chern-Simons action. He used geometric quantization technique. Recall that a quantum state is a probability over the classical space. In this case the classical moduli space consists of unitary flat connections. The Hilbert space is then the space of holomorphic sections of the natural determinant bundle over the moduli space, or non-Abelian theta functions. To have a topological theory one has to show that the quantization is independent of other choices, in this case the choices of complex structures. It is shown that there exists a projective flat connection for the Hilbert space bundle over the space of complex structures [Axelrod-DellaPietra-Witten], [Hitchin]. Knizhik-Zamolodchikov equations are the explicit expression of the projective flat connection for the case of punctured sphere.

In Atiyah's axioms of topological quantum field theory, one also needs to construct morphisms between Hilbert spaces corresponding to cobordisms of manifolds. Witten illustrated that such morphisms are given by the Feynman path integral with Chern-Simons action. He explained relevant background such as Feynman-Kac formula and Feynman diagrams and made it more accessible to mathematicians. There are some interesting works by theoretical phycisists to try to make sense of the perturbation series. It is found that symmetries play a most important role here. After all, Chern-Simons action is the only Lagrangian to enjoy both gauge symmetry and diffeomorphism symmetry. Divergent Feymann diagrams cancelled from each other because of the two symmetries.

We have added some materials to fill details left out and to update some new developments. In Chapter 4 we explained the approach based on representations of mapping class groups [Moore and Seiberg], [Kohno]. Due to time constraint and limitation of my knowledge we omitted many important approaches, e.g. quantum group method [Reshetikhin-Turaev], [Kauffman], Vassiliev invariant [Birman and Lin], [Bar-Natan], [Kontsevich], perturbation series [Axelrod-Singer], among others. We hope interested readers can find some materials from those references.

We also found it very amusing to define knot invariant via string theory or topological sigma models. In Chapter 6 we gave a very brief introduction of this method based on Witten's paper[Witten-string]. There are very interesting conjectures by Vafa [Vafa] about constructing knot invariant from closed string theory recently. We explained localization principle which is usually used to establish mathematical bases for topological sigma models. Localization principle has become a powerful tool in dealing with many problems in topological quantum field theory and in string theory.

It is a pleasure for me to study the notes I took from Witten's lectures. It is not surprising that it takes quite long to absorb and understand some of the materials. I found it very rewarding to study them. I wish to thank E. Witten for answering some of the naive questions from me and for providing a nice Forward. I wish to thank S. S. Chern for his genuine interest and encouragement. I wish to thank Dr. K. K. Phua who encouraged me to write up these notes. I wish to thank B. Abikoff, S. Axelrod, Chen Wei, P. Deligne, Jing Qin, Lin Xiao-Song, Liu Kefeng, H. Verlinde, Paul Yang and Zhang Weiping for many helpful discussions. I wish to thank Lu Jitan and Kim Tan for their professional editorial work. And last but not least I thank my wife, Hou Bo, for her support. Without all of the interest and encouragement above I would not be able to finish this work.

By Sen Hu
November 11, 2000